至臻品质　精准放亮

智慧灯光　还原本色

Idea-Tops 艾特奖

2014艾特奖获奖作品年鉴

国际空间设计大奖艾特奖组委会/编著

湖南科学技术出版社

地址：中国浙江省宁波市高新区聚贤路1299号　电话：0574-27613901　网址：www.clsaself.com

2014 艾特奖获奖作品年鉴
THE AWARD WINNING WORKS OF IDEA-TOPS

图书在版编目（CIP）数据

2014艾特奖获奖作品年鉴/国际空间设计大奖艾特奖组委会编著. —长沙：湖南科学技术出版社，2015.9
ISBN 978-7-5357-8723-1

Ⅰ.①2… Ⅱ.①国… Ⅲ.①建筑设计－世界－2014－年鉴 Ⅳ.①TU206-54

中国版本图书馆CIP数据核字(2015)第172043号

2014艾特奖获奖作品年鉴

编　　著：国际空间设计大奖艾特奖组委会
责任编辑：缪峥嵘
总 策 划：深圳市东方辉煌文化传播有限公司
统筹策划：赵庆祥　朱倩　许蕊
文字翻译：杨小凤　韩笑
出版发行：湖南科学技术出版社
社　　址：长沙市湘雅路276号
　　　　　http://www.hnstp.com
邮购联系：本社直销科　0731-84375808
印　　刷：利丰雅高印刷（深圳）有限公司
　　　　　（印装质量问题请直接与本厂联系）
厂　　址：深圳市南山区南光路1号
邮　　编：518051
出版日期：2015年9月第1版第1次
开　　本：889mm×1194mm/16
印　　张：22.75
书　　号：ISBN 978-7-5357-8723-1
定　　价：498.00元

（版权所有·翻印必究）

CONTENTS 目录

044/	**A** 最佳文化空间设计奖	Best Design Award of Cultural Space
068/	**B** 最佳酒店设计奖	Best Design Award of Hotel
092/	**C** 最佳餐饮空间设计奖	Best Design Award of Dining Space
116/	**D** 最佳娱乐空间设计奖	Best Design Award of Entertainment Space
140/	**E** 最佳展示空间设计奖	Best Design Award of Exhibition Space
164/	**F** 最佳会所设计奖	Best Design Award of Club
188/	**G** 最佳办公空间设计奖	Best Design Award of Office Space
212/	**H** 最佳样板房设计奖	Best Design Award of Show Flat
236/	**I** 最佳别墅豪宅设计奖	Best Design Award of Villa
260/	**J** 最佳陈设艺术设计奖	Best Design Award of Art Display
280/	**K** 最佳公寓设计奖	Best Design Award of Apartment
304/	**L** 最佳商业空间设计奖	Best Design Award of Commercial Space
328/	**M** 最佳光环境艺术设计奖	Best Design Award of Lighting Design

006/ 中国境内国际化程度最高的专业设计大奖IDEA-TOPS艾特奖
Idea-Tops, the Most Internationalized Professional Award for Design in Mainland China

008/ 序言
Preface

014/ 2014 IDEA-TOPS艾特奖评审委员会
Idea-Tops Jury Panel 2014

016/ 设计师们眼中的艾特奖
Idea-Tops in the Eyes of Design Masters

022/ 2014年度艾特奖颁奖盛典
Idea-Tops Award Ceremony 2014

033/ 首届G10设计师峰会
The First G10 Designers' Summit

Idea-Tops, the Most Internationalized Professional Award for Design in Mainland China

中国境内国际化程度最高的专业设计大奖 IDEA-TOPS 艾特奖

国际空间设计大奖——Idea-Tops艾特奖，是中国境内国际化程度最高的专业设计大奖，建基于全球第二大经济体及迅猛发展的设计市场，旨在发掘和表彰最佳设计师和最佳设计作品，打造全球最具思想性和影响力的设计大奖。

International Space Design Award-- Idea-Tops is the most internationalized & professional space design award in China which built upon the world's second-largest economy and flourishing design market. To discover and honor the best designers and design works, Idea-Tops aims to create the most thoughtful and influential interior design award in the world.

艾特奖为推动中西方设计交流搭建了一个沟通协作的平台，高水平的国际级评委、严谨公正的评奖机制及奖项设置……使艾特奖成为中国境内最具国际化和专业性的设计奖项，也成为了世界设计了解中国设计的一个窗口。

Idea-Tops Award is to provide an exchange platform for bridging the Chinese and Western communication and collaboration in the way of design with high level of international judges, rigorous and impartial mechanism along with award settings, offering an unparalleled level of internationalized design award to the world through this important gateway of China.

艾特奖参与者可谓众星云集，包括全球三大设计事务所之一的Gensler设计总监Graeme Scannell、"中国第一高塔"广州塔设计者Mark Hemel，全球酒店设计公司五强、BBG-BBGM建筑与室内设计公司设计董事Robert J.Gdowski，拥有140年历史的国际知名建筑事务所Woods bagot全球总监Rodger Dalling，英国首相官邸——唐宁街10号设计者、BBC苏格兰总部设计师Ross Hunter，希尔顿国际酒店集团主创设计师Martin Hawthornthwaite，曼联俱乐部亚太区首席设计师Mike Atkin，深圳机场T3航站楼设计者Fuksas夫妇，2008年普利兹克奖获奖者、法国当代著名建筑师让·努维尔(Jean Nouvel)等。

The design stars have glorified Idea-Tops Award, including Graeme Scannell, Design Director of Gensler, one of the three biggest global architectural design agencies; Mark Hemel, the designer of Guangzhou Tower—China's tallest tower; Robert J. Gdowski, Design Director of BBG-BBGM, one of the top five global hotel design agencies; Rodger Dalling, the Global Director of Woods Bagot, which has over 140 years of history; Ross Hunter, the designer of 10 Downing Street (Prime Minister's Office in UK) and headquarters of BBC Scotland; Martin Hawthornthwaite, the Chief Designer of Hilton International Hotel Group; Mike Atkin, the Chief Designer of Manchester United in Asia Pacific region; Massimiliano Fuksas and Doriana Fuksas, designers of Terminal 3 for Shenzhen Baoan International Airport; Jean Nouvel, a renowned French architect and the winner of Pritzker Prize 2008.

源于东方，面向世界，Idea-Tops艾特奖崇尚的，是设计师永不枯竭的智慧与前瞻性的革新思想，以及他们在审美和工艺上均显卓越的设计作品。恪守专业、严谨、公平、公正的原则，Idea-Tops艾特奖以较高的专业标准、专业发展、专业责任以及专业沟通来促进设计业的发展，每届艾特奖均邀请全球设计领域资深专家、学者、顶尖建筑师和设计师、知名人士、财经专家、意见领袖以及有影响力媒体担纲评委。

Driven from the East and aims to embrace the world, Idea-Tops recognizes designers' inexhaustible talent and innovative forward-thinking as well as their quality design in terms of aesthetic level and technical artistry. Following the principle of being professional, rigorous and impartial, Idea-Tops is committed to promote the development of design industry by setting high level of professional standards, responsibilities and communications. Distinguished specialists, scholars, top architects and designers, celebrities, financial experts, opinion leaders and influential media-men will be invited to join the jury panel of Idea-Tops every year.

作为表彰建筑和室内设计界杰出人才的重要奖项，获得艾特奖的首肯也就是向全球昭示了精英们在建筑和室内设计界的顶级荣誉。

Being a remarkable award recognizing preeminent architectural and interior designers, Idea-Tops also shows the world the highest glory of its winners in architectural and design industry.

007
IDEA-TOPS
艾特奖

向设计致敬

他们从平凡中来
又从平凡中创造伟大
他们不遵循规则
却重新定义了规则
他们用变革精神
为我们的生命注入色彩
我们尊重并感谢他们
向设计致敬
向设计师致敬

IDEA-TOPS 2014

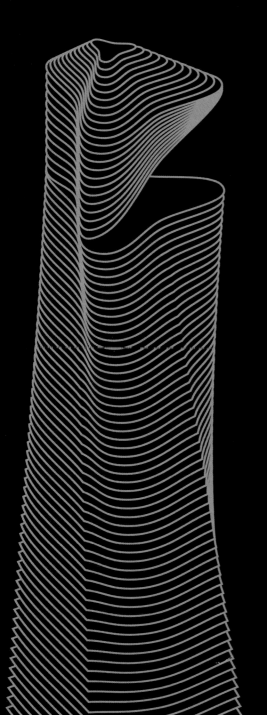

008

IDEA-TOPS
艾特奖

三个成功因子与三个关键要素
Three Success Factors and Three Critical Elements

刘育东

哈佛大学建筑设计博士 / 亚洲大学副校长

Liu Yudong, Ph.D of Harvard University of Architecture, Vice President of Taiwan Asia University

一个设计奖项是否成功，我们能从三个方面来检视。第一是参赛作品的质与量，第二是入围作品对当代设计议题讨论的深入度，第三是得奖作品在国际设计发展中的未来性。我连续2年参与艾特奖，除了看到两届最后胜出的得奖作品外，第二年更参与最终评审，有机会直接观察上述的成功因子。

这一年，再次看到了来自全球不同区域不同国家设计师提交的优秀作品，这些作品受各自文化背景的影响，呈现出不同的视觉感召力。这其中，让人惊喜的是在全球化与在地化的双重趋势下，不少设计师著力于发掘本土设计语言，力求营造出既保留地方历史人文特色，又符合现代人审美情趣与起居需求的活动空间。这种对"传统文化现代化"的深度探索，以及对"变"与"不变"的文化传承思考，是值得整个设计行业期许的努力。

评审过程，每个评委在审评作品时各有侧重，绿色健康、可持续发展、以人为本等都是考量因素。对我个人而言，我重视的是设计情感的传递，因为好的设计师能让空间自己讲话，建筑者、营造者和观看者都有诠释的权力。从我前面提到的三个成功因子看来，经由参赛作品、入围作品、得奖作品构成的艾特奖已经是非常成功的设计类奖项。

素有建筑诺贝尔奖之称的普立兹克建筑奖向来是建筑界的最高荣誉，我在2003年受邀参与颁奖典礼时，曾当面请教大会主席汤姆普立兹克先生："普立兹克奖能成为史上最重要的建筑奖的成功要素是什么？"他告诉我有三个关键，奖项定位要远大、评委邀请要宏观、评审过程要独立。以这三个关键要素看来，艾特奖已具备国际大奖的格局。

设计讲究开创、讲究未来，因而讲究不为今日设计、要为明日而作。因此，一个设计奖项也是主办者、参赛者、评审者共同为"明日设计"投入心血的历程。今天处在一个新旧时代交替的节点，也是新机会的开始。艾特奖为探索新时代的设计方向，以及中国设计、亚洲设计、世界设计融合发展中的交互共生，提供了绝佳的平台，让所有建筑与室内设计领域中勇于创新的人，一同来写"设计的明日历史"。

There are three aspects to tell whether a design award is successful. First, it depends on the quality and quantity of the entries. Second, it depends on the shortlist's exploration of contemporary design topics. Third, it depends on the futurity of its winning works in international design developing process. It's the second year in a row that I am invited to participate in Idea-Tops Award. Apart from the opportunity to witness the award winning works in the past two years, being a final judge in the second year, I was able to take a close look at these success factors.

This year, again, we found good designs from diversified regions and countries of the world, and each of them displayed different visual impact revealing iconic identity of local cultural charisma. What surprised us most was that even under the influence of global and nationwide trends, a good many of designers still dedicated to establish indigenous design language in order to preserve historical characteristics on one hand and to meet modern living standard on another. This in-depth study of the Modernization of Tradition, and the Change and Not Change of traditional culture, are good efforts worthy of the appreciation from the whole design industry.

In the course of the assessment, every judge differed in priorities with regard to environment friendly, sustainable development and people-oriented elements. Personally, I valued sentiment conveyed in the design. It's because a good designer enables space to tell stories, while architects, creators and visitors might have their own translation. Judging from the three success factors I mentioned in the beginning, taking into consideration the entries, shortlist and finalist stage of Idea-Tops, it is already a very successful design award.

The Pritzer Prize is recognized as the highest distinction in architecture community and the Nobel Prize of architecture. When I was invited to attend its award ceremony in 2003, I asked Tom Pritzer who was the president of the organizing committee, "What makes Pritzer Prize the most successful award for architecture?" He told me there were three critical elements. Those were ambitious prize positioning, comprehensive jury panel and independent assessment procedure. Judging from these three elements, Idea-Tops has formed the setup of an international award.

Design features originality and futurity. That is to design for tomorrow, not for today. Therefore, a design award is also a journey in which the organizer, participants and jury make every effort to pursue Design for Tomorrow. We are standing at the interchange of new and old era which is also the beginning of a new opportunity. Idea-Tops provides a great platform for us to explore the design direction of this new era as well as for Chinese Design, Asian Design and Global Design to interact and blend with each other. It invites all innovative people in architectural and interior design industry to compose History of Design Tomorrow.

Giuseppe Giarracca

意大利骑士、建筑师
Knight of Italy, Architect (Pino)

国际空间设计大奖——艾特奖是全球设计圈的一大焦点。它不但表彰了卓越的设计师，也为我们提供了独一无二的分享经验、交流竞技与思维碰撞的机会。来自世界各地的杰出教授、建筑师和设计师齐聚一堂，共同促进、研讨和鉴赏全球设计行业的发展，以及日新月异的生活方式。艾特奖更创造了东西方携手合作和交流的空间。

我很荣幸成为艾特奖的评委之一，得以自始至终亲身体验这场盛典。这段经历非常宝贵，让我有机会接触新理念并评估具有创新性的项目。参赛作品呈现的品质、创新和原创性给我留下了深刻的印象。所有项目显然都对新的设计工艺和原材料做过深入研究。值得一提的是，能够见证这种对创新和卓越设计的嘉奖，我感到很自豪。

在建筑和设计领域，精益求精与持续创新是必须的。就像我们在2014艾特奖盛会中探讨的一样，假如室内设计的品质提高了，生活的品质也会得到改善。选材优良、功能便捷、造型美观的室内环境能让我们的日常生活受益颇多。一件如艺术品般精美的家具也会赋予个人生活不同凡响的意义。

International Space Design Award Idea-Tops is a key focus for the international design community. Idea-Tops has become an honor for exceptional designers to receive, as well as a unique opportunity to share, compare and consider new visions. Outstanding professors, architects and designers from all over the world gather to promote, discuss and admire the development of the Chinese and international design scenes, as well as general improvements in lifestyles. Idea-Tops promotes a space of cooperation and mutual exchange of Western and Eastern points of view.

In the design and architectural domains, China is an emerging country. However, over the last twenty years it has attracted architects and designers from all over the world. The novelties of the Chinese market and its innovative spin have offered an interesting challenge for all professionals who have ventured to China. Thanks to a stimulating and competitive environment, as well as a strong commitment to innovation and progress, the Chinese design and architectural market has flourished. Even though international influences are strong, Chinese designers and architects have managed to develop their own unique style. In other words, the Chinese have taken international ideas and combined them with Chinese culture. Therefore, the Chinese style is a mix of European, US, and Chinese style. Personally, I believe that the Italian style played a major role in the development of the Chinese's approach to architecture and design. Italy places great value on the quality and functionality of its architecture and design. After all, the legacy of the Roman Empire has influenced architects in every nation around the world.

In the fields of architecture and design, there is always room for improvement and renewed creativity. As it has been discussed during Idea-Tops 2014, the quality of life will improve if we improve interior design. Excellent materials and highly functional and beautiful interiors can do a great deal in our daily lives. An art-like and superb piece of furniture can make a difference in the life of an individual.

有中国设计的世界，将大大不同
Chinese Design Will Make a Great Difference in the World

赵庆祥 / 艾特奖组委会执行主席 / 深圳市政协委员
Executive President of Idea-Tops Organizing Committee, Committee Member of CPPCC in Shenzhen

2014年12月16日，国际空间设计大奖——IDEA-TOPS艾特奖14个类别年度最佳设计大奖正式揭晓。经过入围奖评审、提名奖评审以及国际终评三大严谨而权威的评审程序，汇聚了来自世界不同国家和地区的卓越设计力量，这个奖项的获得的确不容易，我们向他们致敬。

近年来，越来越多有高品质需求的房地产开发商、酒店投资方和工程建设甲方，将艾特奖获得者锁定为合作目标，在中国，超过2/3的知名房地产开发商和酒店投资集团已经将艾特奖获得者作为项目合作首选。更多来自世界各地的优质项目投资方、商业集团与艾特奖组织机构进行对接，寻求顶尖级优秀设计师、设计机构开展合作。艾特奖之所以在短短几年获得如此巨大成功，有两个重要原因，一个就是巨大的中国市场吸引了众多境外设计师，他们希望通过艾特奖这个平台，来拓展中国市场。另一个也是最重要的原因，就是艾特奖评奖严谨、奖项权威、专业、公正，获奖设计师代表了当今建筑室内设计的最高水平。

过去五年来，艾特奖先后在四十多个国内外城市进行了全球巡回推广活动，专业影响力、学术影响力、国际影响力日益凸显，国际参与度越来越高。参赛作品每年呈上千件递增，参与者覆盖了英国、美国、意大利、德国、法国、希腊、瑞典、葡萄牙、以色列以及中国大陆、港澳台等35个国家和地区。

2014年艾特奖的揭晓，带给我们新的惊喜，全球空间设计背后的中国力量正在迅猛崛起。琚宾的作品《蜗居27》在不可思议的狭小空间中，展现不可思议的宁静、宽广和无限的设计生命，令人动容。聂剑平的作品《墅家墨娑》则被国际评委们肯定为："出色地保留了历史氛围，内外空间设计的步调非常有聚合力。"特别值得一提的是，"2014年度艾特奖最佳会所空间"获得者洪忠轩，经艾特奖组委会推荐，全案负责迪拜哈利法塔第151层建筑室内设计项目。这是中国设计师首次进军中东市场并担纲迪拜塔顶层室内设计，也是艾特奖向全球推广以来取得的重大成果。艾特奖已经成为诞生明星设计师的舞台。

《2014艾特奖获奖作品年鉴》的出版，是对当代优秀设计师与设计作品的巡礼与致敬，记录了艾特奖获得者的成就高度，也从一个侧面反映了当前全球建筑室内设计的发展水平。我们有理由相信，在艾特奖这个平台上，更多的精彩将会持续，更多的重量级设计师将会脱颖而出。

伴随着设计业的发展，日益坚定自信的中国设计力量将不断刷新人们的期望值。市场终会证明：有中国设计的世界，将大大不同。

On December 16th 2014, International Space Design Award Idea-Tops officially announced the final winners of 14 categories. Going through three major rigorous and authoritative assessment procedures of the entries, shortlist and finalist phases, the design talents from diversified countries and regions are well deserved. We would like to pay tribute to all participants who contributed to this great event!

In recent years, more and more high end real estate developers, hotel investors and project contractors make Idea-Tops winners the target collaborative partners. Over two third of the famous developers and hotel groups have already considered Idea-Tops winners as premier candidates. A great many of developers and business groups from overseas get in touch with Idea-Tops for cooperation with top designers and design companies. There are two major reasons why Idea-Tops becomes so successful in just a few years. One is foreign designers all hope to enter the huge Chinese market through Idea-Tops. Another which is the most important one is that due to its rigorous, authoritative, professional and impartial quality, Idea-Tops represents the highest recognition for interior and architectural designers.

For the past five years, Idea-Tops has organized promotion events in more than 40 cities home and abroad. As it gains more recognition for its professionalism, academic level and international influence, the number of entries increases by almost one thousand every year. Participants were from 35 countries and territories in the world, such as the United Kingdom, the United States, Italy, Germany, France, Greece, Sweden, Portugal, Israel, mainland China, Hong Kong, Macau and Taiwan, etc.

At the award ceremony of Idea-Tops 2014, we were proud to find rising power in Chinese design. The Snail House 27square meters designed by Ju Bin, presents a peaceful and spacious space in such a small apartment. Its unlimited life of design touched everyone. The international jury spoke highly the project Shujia Mosuo of Nie Jianping as, "This is a beautiful design in keeping with the historic context. Both internal and external space has been considered in a cohesive way, resulting in a holistic design." What's more, Hong Zhongxuan, the winner of Best Design Award of Clubs 2014, who was recommended by Idea-Tops Organizing Committee to design the 151st floor of Burj Khalifa in Dubai. Project in Burj Khalifa is the debut of Chinese designer in Middle East market, which is also a significant achievement of Idea-Tops in its internationalization process. In a word, Idea-Tops spawns star designer for tomorrow.

The Award-Winning Works of IDEA-TOPS is a parade and celebration of remarkable contemporary designers and works, reflecting achievement of Idea-Tops winners and global interior design. We believe Idea-Tops will continue to impress the world with more leading designers.

As the design industry keeps on evolving, the growing sophistication of Chinese design will heighten expectation of others. The market will prove: Chinese design will make a great difference in the world.

2014 IDEA-TOPS艾特奖评审委员会

Idea-Tops Jury Panel 2014

1 郑曙旸 2 JULIA MONK
3 DARREN CARTLIDGE
4 刘育东 5 梁景华
6 GIUSEPPE GIARRACCA(PINO)

Idea-Tops in The Eyes of Design Masters

设计师们眼中的艾特奖

12月16日，2014年度艾特奖系列活动在深圳大中华喜来登酒店隆重举行。在室内设计行业迅猛发展，机遇与挑战并存之际，2014艾特奖吸引了全球设计的眼光，共同探讨设计，共话设计未来，形成了强大的磁场力和向心力。一个设计大奖为何能形成如此巨大的影响，掀起行业飓风，艾特奖本身有哪些魅力？在国内外设计大师眼中，艾特奖究竟是什么样的？走近国际顶尖设计大师，感受艾特奖的真实魅力。

On December 16, 2014, the series activities of Idea-Tops took place at Shenzhen Sheraton Futian Hotel. With this moment when opportunities and challenges booming in design industry, Idea-Tops 2014 had captured public's attention worldwide to probe into design and future trends, bringing about dynamic influence. As a top award for design, what makes Idea-Tops so appealing? What Idea-Tops is like in design masters' opinion in domestic or abroad? Let's take a close look and find out the charms of Idea-Tops.

018
IDEA-TOPS
艾特奖

梁景华
香港PAL设计事务所创始人
Patrick Leung, Founder of Hong Kong PAL Design

"越来越多的优秀设计师参与到艾特奖，而且这个奖项已经推广到国际，很多国际知名的设计师都参与其中，奖项的分量越来越重，而且评委也是越来越有分量。艾特奖的影响力越来越大，中国以及国际设计师都慢慢地对艾特奖这个平台形成黏度，所以我认为这个奖项以后会成为不可或缺的国际大奖。"

"I find many outstanding designers participated in the Idea-Tops which has become famous worldwide. As increasing numbers of international famed designers joining this event, the award settings and jury panel are getting more professional too. Both local and foreign designers pay such continuous attention to this award that I think it's going to be one of the most indispensable awards of the world."

梁志天
梁志天设计师有限公司创始人及董事长
Steve Leung, Founder and President of Steve Leung Designers

"我觉得艾特奖是一个非常不错的奖项，因为它把不同领域的设计师包括建筑设计、室内设计、或者产品设计汇聚在一起，然后大家会通过这个活动来增进交流，进行跨界跨领域的交流，促进彼此的成长。"

"The Idea-Tops is an amazing award on my part. Because it brings architects, interior designers and product designers of different fields together, providing an opportunity for them to exchange ideas and facilitate crossover development."

朱宣
深圳市花样年地产集团有限公司执行总裁
Zhu Xuan, CEO of Shenzhen Fantasia Real Estate Group

"整个室内设计和装饰行业，整个产业链的源头就在于原创设计。那么艾特奖最强调的就是原创的艺术价值以及个性的魅力。所以获得艾特奖的设计师都有共同的特点，就是有行业的标杆意义，有原创的价值。同时在审美上是有个性特征的，那么我们根据艾特奖的获得者来选择设计师就会有一定品质的保障，同时也会选择他们的个性，他们的个性也正好是原创设计的价值。"

"Original design is the source for interior design and decoration industry. And what Idea-Tops emphasizes is the fascination of originality and individuality. So the winners of Idea-Tops share those fine qualities in common. Idea-Tops is also a guarantee for quality design and original style."

薛建华
香港卫视常务副总裁兼总编辑
Xue Jianhua, Vice General Manager and Editor in Chief of HKSTV

"艾特奖作为我们深圳设计之都的一个品牌，现在已经有国际影响力了……香港卫视作为艾特奖的一个战略合作伙伴，通过我们香港卫视的国际化平台，使艾特奖的影响力在全球不断地扩展，艾特奖的成长也会为我们香港卫视以文化为灵魂的这样的节目带来更多丰富的内容。"

"Being a brand of Shenzhen--the City of Design, Idea-Tops enjoys an international reputation by receiving entries from greater China, Hong Kong, Taiwan, Macau and dozens of other countries. As Idea-Tops' strategic partner, HKSTV showcases the world this great event with its international coverage and helps it to gain growing influence worldwide. Equally, Idea-Tops' extension means a lot to HKSTV, for our soul is to disseminate programs featuring cultural event like this."

郑 忠
香港郑中设计事务所创始人及董事长
Cheng Chung, Founder and President of Hong Kong Cheng Chung Design

"艾特奖的创立在设计之都的深圳，规模和专业化都定位于国际水准，为设计界评选出具备国际顶尖水平，富含美学意义的作品，同时也为获取殊荣的企业的发展树立了国际品牌。"

"Making a professional design award on an international scale, Idea-Tops was given birth in Shenzhen, the City of Design. It discovers top level and brilliant works in the design community and helps its award winning companies to establish international brand image."

邱德光 019
邱德光设计事务所创始人及总设计师
T.K. Chu, Founder and Design Director of T.K. Chu Design

IDEA-TOPS
艾特奖

"艾特奖已经国际化，所有设计师竞逐这个奖主要目的是希望在国际上曝光，更重要的是在内地曝光，然后慢慢地有更多的接触的机会，我觉得这是很好的现象，那我也希望鼓励设计师来这里参奖，了解一下国际趋势的变化，设计风格的变化，对设计师是有很大帮助的，这就是我的想法。"

"Idea-Tops is already quite internationalized. All the designers hope to gain global exposure through this competition, particularly in mainland China. It is very significant. And I encourage more designers to take part in it and comprehend variable trends worldwide which will help them enormously."

Mark Hemel
"中国第一高塔"广州塔设计者、IBA事务所创始人
Mark Hemel, the designer of China's tallest tower-Guangzhou Tower, Founder of IBA

"在我看来，艾特奖给不同风格的设计师提供交流的可能。所以，东西方的设计师有机会见到对方并了解什么是设计，从而相互影响。总之，交流和互动会有益于设计的发展。"

"I think the Idea-Tops is bringing different type of designers in contact with each other. So both east and west have possibility to meet each other to see what designs are about. Therefore, be influenced by each other. And I hope the designers themselves will benefit from the kind of interaction."

Darren Cartlidge
英国Atkins阿特金斯集团亚太区负责人
Darren Cartlidge, Operation Director of Atkins in Asia-Pacific region

"作为设计师，我们想要世人认可我们在设计过程中施展的技艺。艾特奖让我们更了解彼此，可以借机见到其他志同道合的专业的设计师、客户、研究者和学者等人物，这是很重要的一点。此外，与当今专业的团队和顶级建筑师也有更多交流合作的机会。"

"As a designer, we want to be recognized for the skills that we put into design process. Meeting other like-minded professionals, clients, researchers and everybody in education is always important, it's good to be together. Add to that, you will get more collaborations and key works with today's professional team and architecture."

020
IDEA-TOPS
艾特奖

Julia Monk

美国HOK设计集团全球副总裁
BBG-BBGM联合创办人
Julia Monk, Vice President of HOK,
Co-Founder of BBG-BBGM

"中国的艾特奖表彰了优秀的设计，这对现在和未来都是一个很好的改变。"

"As Idea-Tops is to acknowledge really great design. I think this is a wonderful change for today and for the future."

Giuseppe Giarracca

北京设计中心首席设计师、
意大利骑士、建筑师
Giuseppe Giarracca, Chief Designer of the Beijing Design Center, Knight of Italy, architect

"我认为艾特奖是个很好的概念。它不仅为东西方的设计碰撞提供了交流和理解的平台，也提供了理解年轻和资深设计师、理解客户需求的平台。这种设计的交流给我们设计师带来许多思考。"

"I think Idea-Tops is a great idea. Because it is a good opportunity for us designers to confront between foreign and Chinese design to understand the new design of the current, to understand the problem of the young and old designers, to understand the problem of the design requirement from our clients. The confrontation is our moment of reflection for everyone of our designers."

崔华峰

广州崔华峰设计有限公司创始人、
设计总监
Cui Huafeng, Founder and Design Director of Cui Art Design Studio

"艾特奖是一个关于观点的奖项，整个活动贯穿了很多很有力量的观点，这是非常值得肯定和发扬的。"

"Idea-Tops is all about good ideas. There are a great many of powerful ideas throughout the activities which deserve to be carried on and promoted."

Bert Ghys

比利时皇家建筑学院会员建筑师、KRA设计建筑总监
Bert Ghys, Architect of Royal Academy for Architecture of Belgium Director of KRA Design

"我觉得艾特奖为不同背景的设计师搭建了一个互相交流的平台，交流中对自己的专业技能有所提升，特别是今天的主题演讲和高峰论坛环节，我觉得每位嘉宾都说得非常好，而且现场观众都能够参与进来。"

"Idea-Tops built a platform for designers with different background to communicate with each other, from which promote their professional skills. What impressed us most was the theme speech and the summit forum segment, every honored guest had excellent performance. What's more, the interaction with the audience drove the atmosphere to the peak."

刘育东

哈佛大学建筑设计博士
麻省理工学院共同博士研究
亚洲大学副校长
Liu Yudong, Ph.D of Harvard University of Architecture, Vice President of Taiwan Asia University

"艾特奖从名字上看就是Idea-Tops，就是点子最重要。所以世界各国的点子都能够汇集在一个平台上，一起被交流，被看到，我认为这就是艾特奖最重要的价值。它并不是在单一的文化，单一的设计价值观理念，单一的生活风格里面。而是能够跨越这么多的，不同的文化、气候、风土人情，以及不同的社会。"

"By literal meaning, Idea-Tops is all about the ideas. Ideas in worldwide scope gathering on the same platform, to be appreciated and studied, it is what Idea-Tops matters in my point of view, which combines different cultural background, value and concept, weather, tradition, economy and life style. For that reason, it makes Idea-Tops the most unmissable event to be involved."

Francis Surjaseputra

APSDA亚太设计师联合会会长
Francis Surjaseputra, President of APSDA

"艾特奖系列活动非常丰富生动，让我感受了很多能量。艾特奖大师论坛上提出的问题非常有意义，能引起设计师深度思考。艾特奖也汇集了众多国际设计师争夺大奖，激励设计师共同进步。"

"The Idea-Tops event is very dynamic that I feel a lot of energy. It's a positive energy, people are questioning the negatives, and that's very good. And the Idea-Tops gathered so many international designers to compete and improve."

021
IDEA-TOPS
艾特奖

武石正宣

日本商业环境设计协会国际委员长
Masanobu Takeishi, International Director of Japan Commercial Environmental Design Association

"我是第一次参加艾特奖，感觉十分惊讶，颁奖盛典非常隆重和盛大。G10峰会上的设计理念分享非常特别，感受到中国的设计发展得很快。希望艾特奖越来越好。"

"I was invited to the Idea-Tops for the first time. To my surprise, it was a solemn and grand ceremony. G10 summit was very creative and I found rapid expansion of the Chinese design industry. Wish Idea-Tops become better and better."

苏英姿

美国5+Design深圳公司负责人、绿色建筑设计资格认证中国区域总监
Su Yingzi, Manager of 5+Design in Shenzhen Branch

"今年是我第一次参加艾特奖，我感觉艾特奖组织得非常成熟，整个会场布置考虑得非常周到。艾特奖这个平台能够让室内设计的各个阶层参加进来，这是艾特奖非常成功的表现，我认为这个奖项非常好的一点就是除了针对商业市场范围之外，还考虑到了我们设计上的学术性和思想性，并且关注到现在设计的主要潮流。"

"It is the first time that I joined Idea-Tops, and I think it operated sophisticatedly with well conceived arrangement of the venue. Designers of all levels are welcomed by Idea-Tops which is the key to its success. Besides the considerations for commercial market, it takes academic and innovative thinking into account as well as the key trend in design."

Idea-Tops Award Ceremony 2014
2014年度
艾特奖颁奖盛典

12月16日,全球顶尖设计大师、欧美设计机构高管、亚洲各国设计协会会长、中国大陆及台湾设计院校学术领袖、全国设计师代表、房地产龙头企业高管以及行业媒体记者等近千人齐聚"设计之都"深圳,共同见证了2014 Idea-Tops艾特奖的揭晓。

作为目前中国国际化程度最高的建筑室内设计类奖项,本届艾特奖吸引了来自英国、美国、意大利、德国、法国、希腊、瑞典、葡萄牙、以色列及中国大陆、港澳台等35个国家和地区的建筑室内设计师参赛,共收到设计作品4537件。最终,来自葡萄牙、以色列、中国台湾、深圳、广州、杭州、成都等地的优秀设计师分获本届艾特奖的13项大奖。

中央电视台、香港卫视、深圳卫视、南方卫视、广东电视台、凤凰网、中国网、中国新闻社、人民网、《南方都市报》、《羊城晚报》、《南方日报》、《深圳特区报》、《深圳商报》、《广州日报》等近百家主流媒体对本届艾特奖进行了报道,在社会各界引起广泛关注和强烈反响。

On December 16th, 2014, with the witness of top designers around the world, executive officers of European and American design agencies, presidents of various Design Associations in Asia, academic leaders from design institutes in mainland China and Taiwan, designer representatives from around the country, executive officers of leading real estate companies and professional media, Idea-Tops was unveiled in Shenzhen, the City of Design.

Being the most internationalized architectural and interior design award in China, Idea-Tops 2014 reinforced the continued success by 4537 pieces of entries across 35 countries and territories over the world, such as the United Kingdom, the United States, Italy, Germany, France, Greece, Sweden, Portugal, Israel, mainland China, Hong Kong, Macau and Taiwan. Designers from Portugal, Israel, Taiwan, Shenzhen, Guangzhou, Hangzhou, and Chengdu became the final winners throughout 13 categories of 2014.

About a hundred of mainstream media broadcasted this grand ceremony, including CCTV, HKSTV, SZTV, TVS, GDTV, Phoenix TV, China.org.cn, China News Service, People.cn, Southern Metropolis Daily, Yangcheng Evening News, Southern Daily, Shenzhen Special Zone Daily, Shenzhen Economic Daily and Guangzhou Daily. It had attracted great attention and wide interests from all walks of life.

026
IDEA-TOPS
艾特奖

028
IDEA-TOPS
艾特奖

029
IDEA-TOPS
艾特奖

031
IDEA-TOPS
艾特奖

The First G10 Designers' Summit

首届G10设计师峰会

12月16日上午，2014首届G10设计师峰会在深圳大中华喜来登酒店隆重举行
On the morning of December 16th, 2014, the grand opening of the first G10 Designers' Summit took place at Sheraton Shenzhen Futian Hotel.

034

IDEA-TOPS
艾特奖

G10设计师峰会,由Idea-Tops艾特奖组委会倡导并发起,定位每年一届,于国际空间设计大奖——艾特奖颁奖盛典期间举行。G10(英文GROUP10),是10个研讨小组的简称。为达到深度探讨交流之目的,每届G10设计师峰会限邀100位不同地区、不同文化背景且极具代表性的设计师参与,并由业界极具影响力的设计师担纲峰会主席及主持人。峰会主题则由组委会结合当下世界设计业的发展格局、现状、问题及趋势,进行广泛征集并最终筛选确定。G10设计师峰会旨在为每一位参会设计师提供一个平等的发声机会、思想碰撞以及交流合作的平台,以此引领设计业的发展,推动世界设计产业的共同进步。依托中国境内国际化程度最高的设计大奖——艾特奖,G10设计师峰会备受业界关注。

首届G10设计师峰会堪称一次设计界思想的大碰撞,本届峰会由2014艾特奖推广大使、国际著名设计师梁景华博士主持,以"蜕变中的东方"为主题展开,围绕10个相关话题进行讨论,共有来自欧美、亚太、中国大陆和中国台湾、香港等10个代表团参与,参与者既有国际知名资深设计师代表,也有新锐设计师,还包括设计学术界权威代表。

036

IDEA-TOPS
艾特奖

On the morning of December 16th, 2014, the grand opening of the first G10 Designers' Summit took place at Sheraton Shenzhen Futian Hotel.

G10 refers to Group 10 which was initiated by the Idea-Tops Organizing Committee, is schemed to hold annually during the Idea-Tops Grand Ceremony. To achieve the goal of profound communication, each year G10 will invite 100 designers on behalf of diversified regions and cultural backgrounds to attend the meeting which is hosted by a renowned designer, discussing multiple perspectives over the current circumstances and future trends of design, at any rate the specific subject will be selected by the Committee. G10 provides each designer opportunity to express one's inner voice leading to the corporation and development of global design. Thanks to the Idea-Tops Award, the most internationalized award for space design in China, G10 was given tremendous attention from the design community.

The first G10 Designers' Summit can be termed as a collision of design ideas. Summit 2014 was hosted by Patrick Leung, a world renowned designer, and 10 groups of delegacy debated about Metamorphosis of the East. The group members came from the Europe and America, Asia Pacific, Taiwan, Shenzhen and Hong Kong, South, East, North and West China, including world celebrated and rising designers together with academic authorities.

037
IDEA-TOPS
艾特奖

040
IDEA-TOPS
艾特奖

IDEA-TOPS
艾特奖

044
Best Design Award of Cultural Space
最佳文化空间设计奖

艾特奖
最佳文化空间设计大奖
BEST DESIGN AWARD OF CULTURAL SPACE

INTERNATIONAL SPACE DESIGN AWARD

获奖者
李军（中国·成都）

获奖项目/Winning Project
婴儿园艾玛仕幼儿园/ Aimashi Kindergarten

046

IDEA-TOPS
艾特奖

获奖项目/Winning Project

婴儿园艾玛仕幼儿园
Aimashi Kindergarten

设计说明/ Design Illustration

这是一个属于童话的王国！入口的左边是星球造型的花池，种上各色的植物代表了不同行星，种了树的花池也可供孩子攀爬玩耍。右边是五颜六色的太空涂鸦墙，小朋友可以用粉笔在上面任意创作并大胆展示。入口的路面有透水砖、沙子、石头、草地等材料，是为了让孩子感受丰富的材质。

进入门厅仿佛进入了小动物的树下洞穴，吊顶全部采用弧形木质造型像是走在大树根下的世界，三个动物造型门洞——小猫、小熊、小兔的剪影更像是小动物穿梭留下的轮廓。左边是更新园所食谱与动态的电脑，家长可以自带U盘拷走以便了解小朋友的成长。右边则是园长办公室，让家长在进入幼儿园的第一时间便可接触到园长，省去了繁琐的奔波。

早教中心地面全部软包，为步伐还不稳当的低龄段小朋友提供了安全保障。当小朋友在波波球池里撒欢的时候，家长们则可以惬意地坐在角落弧形造型的凳子上歇息。

在走廊中穿行，抬头一看仿佛是树根的缝隙透下天光。每间教室的外面都有吸音板做的帆船或城堡的剪影，可供班级粘贴动态信息和照片活动的同时，不破坏乳胶漆墙面和整体美观。

教室内部全是采用原木的家具，房间的外部靠近窗户还有地台，小朋友可以排排坐在台阶上听故事！把做好的手工挂在长满树枝的柱子上。

教室卫生间细致贴合小朋友的使用需求，整个墙面用吸味吸潮的天然硅藻泥涂刷，隔断上利用园所专属的LOGO分男女，各种小动物造型的水龙头，还有适合小朋友个子的镜子。婴儿班卫生间还特别增设浴缸，方便了尴尬意外时老师为小朋友冲洗，可收拉的楼梯也节约了并不宽敞的卫生间空间。

楼上的木工房，隔断是用小剪刀、小扳手等造型组合，这样可以教小朋友将各类工具分别放入合适的位置。舞蹈室，整个弧形的顶全是蜂巢造型，置身于蜂巢中翩翩起舞会不会滴下甜蜜的蜂蜜呢！

蹦跶累了，去图书室坐坐。高高的台阶，小朋友可以任意坐着或者趴着看书、看投影仪投放的影片，楼梯中间的滑滑梯刺激又好玩，下面还有软软的垫子为小朋友保驾护航。

集市扮演区有——西餐厅、果子铺、小吃店、美容院、医院。小朋友的梦想是什么呢？在这里他的所有职业梦想都可以逐一实现，他们在这里和小朋友买卖或交换物品忙得不亦乐乎。

This is a fairyland for children. The doorway is covered by bricks, sand, pebbles and grass that they can feel and touch the nature. On entering the gate, it is a space concept garden which features colorful plant in name of different planets and a graffiti wall for chalk painting.

The hallway is as fascinating as the rabbit hole in the fairy tale. The curved wooden ceiling is like a big tree embracing the entire space. Three animal shaped doors induce them to take an inside look for what's hidden behind. On the left sits the self-service area for parents to check and download the latest data as children grow, while on the right sits the Principal's office of kindergarten, for the parents' get hold of the headmaster in the first place.

The entire floor of early education center is made by soft material catering for the toddlers. When they are playing in the ball pool, the circular chairs at the corner can be an agreeable place for parents to have a rest.

Natural lighting is introduced in the corridor. Outside of each classroom, boat or castle shaped silhouette made of acoustic boards are put for bulletin and memento update.

All the furniture in each classroom is made of wood. There are a couple of steps by windowsill where children can sit and listen to their favorite stories or to exhibit their DIY works on the pillar.

Looking into the bathroom, the wall is applied with smell and moisture absorbent paint with exclusive girl and boy LOGO on the door. Animal shaped taps, children's height tall mirror and handy pull-out staircase are all designed for the small space. Moreover, bath tub is put up for toddlers just in case.

Partition wall of the DIY room upstairs is combined of hand tool styled drawers to help children assort them at the right place. As for the dance room, hive pattern fills the whole arc ceiling as if dancing among the flowers.

When they want to take a break, books and films in the library may meet the needs. While the slide in the middle of stairs offers active kids some exciting and recreation time with safeguard cushion beneath.

In the role play street, they will find restaurant, candy shop, SPA and hospital for trading and exchanging commodity where every dream of them will come true.

051
IDEA-TOPS
艾特奖

NOMINEE FOR BEST DESIGN AWARD OF CULTURAL SPACE
最佳文化空间设计提名奖

谷鹏（中国·济南）

获奖项目/Winning Project

原生之爱——手工布艺体验馆
Fabric Art Home

设计说明/ Design Illustration

本案通过对传统民艺的梳理和解读、感知与传承，并与当代空间设计语言搭接与实验，演绎手工布艺从原生棉花至引绪成纱再到织造成布的过程。进而思考传统民艺发展的方向，如何协调保护与发展，传承与衍生。空间设计秉持对传统民艺保护与再造的理念与设计师尝试寻找传统手工转化当代设计的路径，探寻传统手工民艺发展无限新的可能。

项目说明：
1.项目地址：山东省滨州市博兴县开发区
2.项目面积：130m²。
3.设计说明：整个空间采用双曲面设计手法，在蓝色背景下，凸显纺线的艺术张力。
4.主要材料及工艺说明：
材料：棉花、棉绳、手工棉布。
工艺：以无印蓝布为地，以手工棉布为帘，以原生棉花为饰。穿线悬挂、编绳结网以呈空间。
史料："比之桑蚕，无采养之劳，有必收之效。垺之枲苎，免绩缉之功，得御寒之益，可谓不麻而布，不茧而絮"—— 南宋·陈旉《农书》

In this experimental project, traditional folk handcrafts are approached by contemporary design. We may have an idea of how fabric is processed from cotton to cloth and contemplate where folk craftsmanship will end up, the possibility can be beyond imagination. The designer dreams to preserve that artistry and tries to translate it into modern language.

Details:
1. Location: Boxing, Binzhou, Shandong Province
2. Area: 130m²
3. Idea: designed in double curved surface, textile stands out from the blue background
4. Main materials: cotton, cotton cord, handmade fabric

053
IDEA-TOPS
艾特奖

055
IDEA-TOPS
艾特奖

NOMINEE FOR BEST DESIGN AWARD OF CULTURAL SPACE
最佳文化空间设计提名奖

德建建筑设计咨询(上海)有限公司

获奖项目/Winning Project

上海德法学校图书馆
Eurocampus New Library Shanghai

设计说明/ Design Illustration

德法学校是德国学校与法国学校在上海的联合校区,为大约2000名学生提供高质量的学习环境。10年前,德建受委托负责管理原校区的建造,如今,又在校区庭院内新设了图书馆和学习中心。作为上海德法学校的新中心,特殊的外立面设计象征学习的重要性,并揭示积累知识进而放飞想象力乃读书之要义。因此,依设计定制的"展翅飞翔的书本"不仅向人们展示了光与影的概念,调节日光的渗透,营造柔和微妙的室内阴影效果,同时也以强烈的文化象征喻义阅读触发的无限幻想和创造力。建筑位于底层架空柱之上,以遮顶的开放运动场地代替寻常的设有封闭房间的一层,二、三层则实现了小学部和中学部的连通。新图书馆的玻璃幕墙上安装了特别定制的书形遮阳板。

'Eurocampus', this joint campus of the French School and the German School in Shanghai is providing perfect learning environments for up to 2'000 students. Virtuarch has been in charge of the original campus project 10 years ago and has now inserted a new library and learning center into the courtyard of the school.

As the new heart of the 'Eurocampus', the special façade design symbolizes the importance of learning and the principle to 'have thoughts fly' on the basis of knowledge accumulated in school. Therefore, the custom-made 'flying books' do not only provide an interesting light and shading concept for controlling sunlight penetration, while functioning as decoration item due to its delicate interior shadow effect, but are also a strong cultural reference for the fantasy and ingenuity reading is triggering. The building is placed on 'pilotis', allowing for generous covered playground on the ground floor level and linking the different wings of the Elementary and Secondary School on the second and third floor levels.

NOMINEE FOR BEST DESIGN AWARD OF CULTURAL SPACE
最佳文化空间设计提名奖

UArchitects（荷兰）

获奖项目/Winning Project

T Hofke学校
School T Hofke

设计说明/ Design Illustration

这所小学的情形有些特殊，因为它容纳了5种不同需要的人在其共事，并共享教育、护理、运动和社区设施。它进一步划分为16个教室，其中3组用于日间照料，1组用于课外活动，1个幼儿园和1个体育馆。此外，这栋建筑还与附近的居民保持联系互动，提供剩余的会议室供集会活动。

建筑的外表沿用了前学校的白色基调，白色砖块和木条的竖纹肌理是受到树皮的启发，这也与整个设计过程息息相关。与此同时，对以前建筑的纪念是很重要的。砖块全部是就近取材，没有从其他地区或国家进口，减少了对资源和环境的浪费。

学校中间的开放区域是源于之前的建筑设计，这里不但是左右两边建筑的连接处，还是一个可灵活使用的公共空间，为学校内的人员有更多的接触机会。

在新设计之中，原学校的影子依稀可循。一楼宽阔的楼梯直达大堂入口，建筑中央凹面的设计是为了突出主入口。主楼梯运用了多种透视和横断面的表现，完成了环境由外向内的转移。

从学校的外窗可以发现，随着学生年龄的增大，每层楼窗户的高度也增加了。另外，为了体现教室的特性，每个教室窗户的宽度也不一样。由于这一特点，室内的观景体验也不尽相同。

学校空间恰如其分地表现了个性与教学的需要。室内也处处显示了不同年龄群体的有效次序，例如初中和高年级的区别。因此，建筑充满了方位感，不需要标识系统就能让学生轻易地找到路。设计师将每个独立的学习空间与外面的宽阔走廊融入一起。随着数码学习的增多，教室内的数码教学也越来越普遍。为了利用更多的自然光线，设计师专门放置了挂书包和外衣的橱柜，而不像荷兰的其他学校一样，把衣物挂在过道中。如此以来，走廊中的气氛更加和缓，视野更加开阔。

This is a special primary school because there are 5 different groups of users who work together in one building, and use the facilities together for education, care, sports and neighborhood facilities. It consists of 16 classrooms, 3 groups for daycare, 1 group for extra-curricular activities, 1 nursery school/kindergarten and one gym. The building has a social connection to the neighbourhood and also has rooms available for associations that want to use the extra consultation and meeting rooms.

The white colour of the brick facades in this project is a direct reference to the same color of the previous school. The bricks were laid vertically in order to approach the tree bark (texture) effect of the monumental trees which also influenced the design process. The texture and color of the facade refer to and connect to the context. The bricks are produced in the vicinity of the site in order to reduce the effects on the environment, so that no bricks are imported from another region or country.

The spatial core's open space of the building in the middle of the school is derived from the existing school and also forms the link where the school's different users meet. This is literally the connection between the left and right side of users and the core of the building contains the flexible spaces for general use.

The spatial characteristics of the existing school are recalled in the new design. At the entrance to the hall, a open space is created with a wide staircase leading to the first floor. An indentation at the centre of the building emphasises this main entrance. The transition from outside to inside is made via the school's main staircase with all kinds of perspectives and cross-sections.

The different age groups are referred to in the height of the window openings which become increasingly higher with each floor, as the age group gets older. The width of the windows is different in each classroom in order to emphasize the individuality of each classroom and each child. Differentiation from the inside gives each room a different experience and view to the outside.

Individuality and education also combine perfectly with space. There are lines of sight everywhere throughout the building in order to connect different age groups such as the junior school, middle school and upper school in an orderly way. Orientation and an overview of the building make it easier for the children to find their way through the building which means we don't need to develop signage for this purpose. We decided to combine individual learning spaces outside the classrooms with a view with the classrooms, and connect them to the wide corridors. Digital learning spaces deepen the curriculum and digital teaching also takes place in the classrooms. Natural daylight is taken into account here. We decided not to use the corridors to hang up coats, as is the custom in the Netherlands, but rather to set aside special cupboards and spaces to hang up bags and coats, so that an atmosphere of calm and perspective reigns in the corridors.

063
IDEA-TOPS
艾特奖

NOMINEE FOR BEST DESIGN AWARD OF CULTURAL SPACE
最佳文化空间设计提名奖

何建锋、彭妮妮、邹志刚
（中国·东莞）

获奖项目/Winning Project

《墨集》中国东莞图书馆书友会
Moji Dongguan Book Club

设计说明/ Design Illustration

这个空间是210平方米左右，定位为小型主题演讲或书友聚会使用。但要求尽量能让空间互动，尽量也可以在控制整个预算的同时把该有的气氛巧妙地表现出来。

面对挑战，我们从空间周围的环境出发，记得设计前期，我们站在空间的每个角度进行设计实况模拟体验，力求人停留的每一个视角都与自然和环境为友。

于是按照设想，我们开始寻找与推敲立面的设计形态（因为整馆建筑形态是现代体构成式的），我们也从内外协调性考虑，让里外能相互融合。这是一个关于景、关于光、关于人的变化与感知的空间。

The 210 sq.m space features small lecture space or book lovers gathering spot. We aim at creating space that is interactive and apt to many occasions within the budget.

Based on environment-friendly idea, during the conceptual design, we treat this space from the point of view of its end user.

We held up to its architectural configuration, hoping to establish consistency from its external to the internal. Accordingly, landscape, light and people are what this space is all about.

IDEA-TOPS
艾特奖

艾特奖
最佳酒店设计大奖
BEST DESIGN AWARD OF HOTEL

IDEA TOPS

INTERNATIONAL SPACE DESIGN AWARD

获奖者
聂剑平 (中国·深圳)

获奖项目/Winning Project
墅家·墨娑 / Shujia Mosuo

070

IDEA-TOPS
艾特奖

获奖项目/Winning Project

墅家•墨娑
Shujia Mosuo

设计说明/Design Illustration

西冲村位于中国最美乡村婺源，相传为西施终老之所。走入村口，苍翠的巨大古树静静矗立，透过茂密的枝叶向村内望去，青山绿水粉墙黛瓦，墅家墨娑西冲院在影影中展现开来。谁曾想到，这处舒适的乡间度假居所是由一栋近200年历史的清朝老宅和一座破败不堪的家祠改造而来，设计师改写了古宅的前世今生。

徽派民居受布局和采光的限制，容易带给人阴冷逼仄的感觉，设计中需要解决的一大问题是如何满足现代人的住宿功能需求，让客人在感受老宅岁月气息的同时有一份放松舒适的居住体验。如何在恢复古建筑的同时有所创新，以适应现代人的审美需求？设计师围绕这个问题做了大量的工作。传统徽州老宅最大的特点是有天井无院落，视觉感官比较阴暗难以久居，设计师利用家祠前的空地加建一栋由一层咖啡厅和二层水景房构成的两层小楼，家祠与小楼自然形成了一处有回廊的院落，使空间变得更有层次感。所有古建筑天井及公共部分完全按照老宅原样恢复如旧，而客房室内沿外墙一侧保留了原样，新隔墙均为白色石膏板面刷涂料，地板刻意挑选了带节疤柞木，原有木结构体均保持原样，自然而不露痕迹地将新与旧完美融合。

室内色彩基本以黑白灰为主，局部使用跳跃的红色、绿色、黄色，使得空间不显沉闷，充满了现代时尚的气息。家具则大部分根据当地徽州家具款式做了简化设计，上色则从法国新古典家具中吸收灵感，上了3种不同灰色。同时为了让建筑与乡村生活融为一体，老宅前开挖了一处水塘，将原本完全幽闭的徽州民居改造成一个远山、近水、休闲平台、咖啡厅、祠堂内外交融相互呼应的休闲空间，古典美与现代美和谐共生，完美地展现在人们面前。

Xichong is a town of Wuyuan, where is said to be the most beautiful countryside in China. Leafy ancient tree towering inside the green hills and blue water, behind white walls and grey roof tiles there is the Shujia Mosuo inn. To your great surprise, this countryside inn is reformed from a 200-year old chapel date back to Qing Dynasty.

Due to its poor layout and lighting of Anhui architectural style, the designer prioritizes its cold and cramped conditions to meet the needs of modern dwellers, at the same time to preserve the sense of old age.

To improve the lighting issue, the designer built another 2-storey house consisting of a café shop and waterscape rooms. It is connected with the old chapel by a corridor, creating a sense of depth. All the patio, public space and wood structure of the old chapel remain the same as before. As for the changes indoors, partitions are painted white and floor is chosen to be knotted oak. The new and the old are brought into one seamless unity.

White, grey and black compose its general color, whereas to add modern look in the hotel, it is mixed with red, green and yellow. Furnishings which given new life to takes the local fitting style as prototype but color matching like neo classical furniture of France. Moreover, a pond was dug for the inn to fit into the countryside landscape. The enclosed chapel changes into a leisure space sitting on mountain, pond and café shop. No matter the beauty is of classic or modern, there is no barrier in between whatsoever.

071
IDEA-TOPS
艾特奖

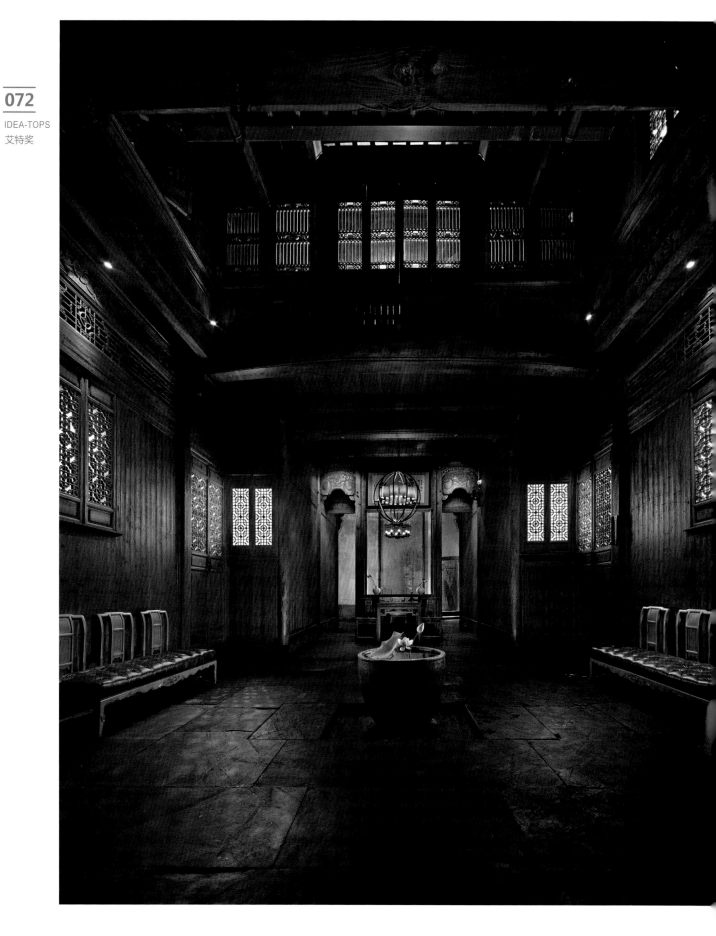

IDEA-TOPS
艾特奖

NOMINEE FOR BEST DESIGN AWARD OF HOTEL
最佳酒店设计提名奖

YANG杨邦胜酒店设计集团
（中国·深圳）

获奖项目/Winning Project

深圳回酒店
Hui Hotel Shenzhen

设计说明/ Design Illustration

回酒店（HUI HOTEL），深圳首家新东方设计精品酒店，由多多集团与YANG杨邦胜酒店设计集团共同斥资并倾力打造。坐落于深圳市千米绿化中心公园旁，紧邻华强北，由旧厂房改造而成，于2014年7月11日正式开业。

"回"作为中国传统文化中最具代表性的文字，它的古文字型是一个水流回旋的漩涡状，寓意旋转。《荀子》有云：水深而回。在中国人传统的人文情怀中，"回"也是人们内心最基本的渴求。无论出发多久，都希望能够回家、回归最初的自我。所以酒店以"回"为名，并以"回"作为设计灵感，让都市生活回归原点，自然回归都市，爱回到家中，让一切感知在艺术空间中得以复苏。

"回"归自然的东方美学意境
酒店整体设计以新东方文化元素为主，并通过中西组合的家具、陈设以及中国当代艺术品的巧妙装饰，呈现出静谧自然的中国东方美学气质。大堂设计中，整面绿色墙植与一字排开的鸟笼，让空间去繁为简，加上清脆悠扬的鸟叫，让人有"蝉噪林逾静，鸟鸣山更幽"之感。酒店整体用色和谐统一，结合用心调试的每一束光源，让空间散发出宁静优雅的文化气质。
空间中一步一景，鲜活翠绿的墙面绿植、精心挑选的黑松、低调简单的哑光石材、波光粼粼的顶楼水面、质朴自然的木面材料……将自然界神秘悠远的天地灵气带到酒店空间中，让人仿若置身旷阔林间。六楼至顶层的部分空间被打通，构建下沉式的室内庭院景观，并引入时下流行的书吧，让宾客在自然雅致的环境中获得身与心灵的双重享受。
"回"归都市里的静谧之家
空间中随处可见的中式柜台、鸟笼、水缸、算盘、灯笼等极具代表性的新中式元素，将深沉、内敛的中国传统文化精髓表现得淋漓尽致。"粤色"中餐厅设计得巧妙独到，天花使用木梁结构处理，极具岭南建筑特色。精心栽种的美人蕉与窗外水景结合，颇有"雨打芭蕉"的诗意。空间中一幅幅熟悉的场景，唤起每一位宾客内心对于"家"的眷恋。
"回"归以人为本的贴心服务
酒店服务注重"以人为本"，除传统的酒店服务外，酒店还提供私人订制贴身"管家式服务"：移动入住、个性化商务、旅游、秘书及私人代订服务等；首创业内"看到即买到"创新服务——所有展出艺术品、家私设备均为独家定制，看中即可购买。客房提供独立衣帽间、大角度变换的电视机、360°旋转衣柜、灯光一键控制系统等完美配套设置，用心呵护每一个细节。
HUI HOTEL将中国传统文化中最具代表的"回"深挖到极致，让心灵、文化、自然在酒店空间中得到回归。它既是一次东方文化国际表达的呈现，也是对"中式风格酒店"的一次最新探索与诠释，树立起中国当代高端奢华精品酒店的标杆。

Hui Hotel is the first boutique art hotel in Shenzhen. Located in the most prosperous business circle, Hui Hotel used to be an old factory. Shenzhen Central Park nearby provides free natural landscape for this Hui Hotel. Although the total investment of the hotel only takes up less than 1/3 of the five-star hotel, it is the most distinctive hotel in this area.

The design of the hotel focuses on new oriental elements and creates peaceful and natural Chinese aesthetic effect through a mix of Chinese and western furniture, as well as the smart decoration of displays and China's contemporary artworks. In the lobby, the whole piece of living wall and bird cages in a row make the space simpler. In addition, Hui Hotel seems to be more quiet while the birds are singing. The hotel applies harmonious colors and carefully adjusted lighting, which makes the space full of peaceful and elegant cultural ambiance.

The most representative new Chinese elements such as Chinese counters, bird cages, water vats, abacus, lanterns etc. can be seen everywhere in the space, revealing the profound Chinese culture essence incisively and vividly. The design of Chinese restaurant "Yuese" is ingenious. Its ceiling is treated with wood beams, making it filled with Lingnan architectural features. The combination of the canna and water features outside the window is like raindrops rattling on the leaves of bananas. By applying the lodgepole pine, living wall, water surface, honed stone, wood finish etc., and the designer brings the mysterious and remote essence from nature to the hotel, making you feel like staying in a vast forest.

Some space from 6th floor to top floor is open, forming a sunken garden. Meanwhile, a library which is popular nowadays is introduced, allowing guests to relax in such natural and elegant environment. Hui Hotel explores the most representative point in China's traditional culture and enables spirit, culture and nature to get back to the hotel space. It is a wholly new exploration and interpretation to "Chinese style hotels" and sets up a benchmark for China's modern luxury boutique hotels.

078

IDEA-TOPS
艾特奖

079

IDEA-TOPS
艾特奖

NOMINEE FOR BEST DESIGN AWARD OF HOTEL
最佳酒店设计提名奖

BK Architecture（以色列）

获奖项目/Winning Project
法兰克福温德姆大酒店
Wyndham Grand Frankfurt Hotel

设计说明/ Design Illustration

法兰克福温德姆大酒店的定位是打造一个当地居民、时尚人士、周末旅行者的都市交流平台。我们倾听这些目标群体的意愿，并界定了这间酒店的概念。

现今的旅行者更倾向于搜索一种具有真情实感的酒店，即在飞逝的时光里抓住每一个城市富有意义的、难忘的瞬间。而我们的酒店就充满了这样的情感。

其整体是都市风格的设计，体现了这个城市的特性：包容性、务实、粗犷、令人出乎意料、真切、酷和友善。它有异于其他法兰克福的奢华酒店，而又不失分毫魅力，是这片乐土上的有力竞争者。我们将都市融入公共空间的设计当中，超越了人们对传统商务酒店的认识，更注重融入性的诉求。

我们期待客人和当地居民也会爱上这一全新的体验，在这里有着最好的酒吧和餐厅，是休闲集会的焦点，而凑巧的是楼上还提供了客房。

The Wyndham Grand Frankfurt is designed as an urban platform and a social hub for the locals, the suitors, the trend seekers and the weekenders. We listened to these groups' desires and expectations as they are active participants in defining our hotel concept.

Travelers today are looking for a charged, authentic, under the radar experience. They want to feel that each of their fleeting moments in any city is meaningful, well exploited and unforgettable. Our hotel steps in exactly at this point.

Its design is urban oriented. Our design draws its DNA from the city public venues: inclusive, down to hearth, rough, unexpected, authentic, cool and friendly, to use some of its values. It proposes an alternative to Frankfurt's posh hotels without losing a drop of its ability to serve as a highly competent player in this playground. We extended the city into our public spaces with the intention to transcend the classical perception of business hotel and move over to an updated mode of engagements.

We expect from our guests and locals to fall for this new alternative and while supported by the best bar and restaurant venues, turn it into an iconic social hub which just happens to have hotel rooms above.

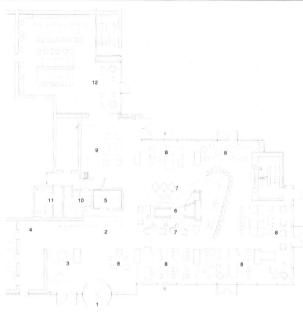

1 entrance 2 reception 3 library 4 elevator lobby 5 back office 6 existing shafts
7 fire place / bar / computer center 8 seating arrengement 9 buffet / conference area
10 women restroom 11 men restroom 12 extension breakfast area

083
IDEA-TOPS
艾特奖

NOMINEE FOR BEST DESIGN AWARD OF HOTEL
最佳酒店设计提名奖

WAF Architekten(德国)

获奖项目/Winning Project

Generator 青年旅社
Generator Hostel

设计说明/ Design Illustration

Generator青年旅社是一个以设计为导向的旅馆品牌,一般所处地理位置极佳。2010年WAF公司被选为这个项目的设计者,项目于2013年底竣工。

这里原本是一个9层的综合办公楼,外表颇具历史感,早在1900年代便已存在,在那时被用作工业仓库,1990年代曾翻新过,地下停车场还可容纳30辆小车。

建筑的里面几乎褪为原本的混凝土颜色,只有楼梯和建筑正面未受太多影响。通往地下停车场的坡道被拆除,多余出来的空间改成了一楼的酒吧。为了让车辆和货物到达地下室,还安装了汽车升降机。

连室内的配色方案都受到灰色陈旧工业建筑的启发:地板、墙面和天花运用了中性的灰色调,不过,要排除色彩鲜明的"主题墙壁"。走廊和房间中的色彩是为了建筑的方位感着想,更重要的是为了让内饰更加鲜活以及突出过道的设计。

Z字形走廊是原建筑的一个功能性设计,而今漫步其中更增加了美妙的视觉体验。

为了最大限度地利用有效空间,洗手间被设计成湿衣室。同时尽可能地避免视觉的障碍物:洗发水的瓶子被嵌入龛穴中,花洒则选用嵌壁式的。

Generator青年旅社的标志性特征就是"超大图形"——在墙上运用大面积的图案装饰。因此WAF公司决定请来当地的街头艺术家Thierry Noir,他是其中最早的一个在柏林墙上作画的艺术家而被众人所熟知。

WAF公司获得Thierry Noir的许可,以他著名的"头像"主题涂鸦作为旅馆走廊独特的壁纸设计。一次性印制了400平方米的壁纸,其中每一个头像都是不同的。各种主题颜色也代表了不同的楼层。此外,Generator青年旅社特有的"潮G"标志,被光雕在浴室内的水龙头和房间门号上。

一楼的办公室循环再利用了原来办公楼中的灯饰,是十分环保的举措。

Generator Hostels is a rapidly expanding brand of design-led hostels in great locations. WAF Architekten were chosen to lead the design from inception in 2010 to completion in 2013.

The existing 9-storey office building complex was bought. It consisted of an exciting historic part to the back of the site, built in the 1900s as an industrial warehouse and a newer 1990s part to the street. There was also a ramp to underground parking for 30 cars.

The building was completely stripped back to the concrete structure. Only the staircases and facades remained largely untouched. The ramp to the underground parking area was demolished to make more space available for a bar on the ground floor. To get vehicles to the basement service and storage areas, and the disabled parking spaces, a car lift was installed.

Even the colour scheme was inspired by the greyness of old industrial buildings: Neutral background grey colours are used for the floors, walls and ceilings, but there is always a "feature wall" with a splash of colour. The flashes of colour in the hallways and rooms help orientation in the building but more importantly create dramatic interiors making even the corridors memorable spaces.

The zigzag walls in the hallways came about as a solution to a technical problem (the location of vertical services) but add a dynamic to the experience of walking down the corridor.

Guest bathrooms are designed as wet rooms to maximise use of available space. To remove visual clutter, as much as possible was recessed into the walls – niches for shampoo bottles, recessed shower taps, etc.

A Generator Hostel trade mark is "super graphics"- large scale wall graphics. WAF Architects decided to be pro-active and involve local street artist, Thierry Noir, known as one of the first artists to paint on the Berlin Wall.

WAF obtained permission from Thierry Noir to use his famous "heads" motif and designed unique wallpaper for hallways. A one-off run of 400m² wallpaper was printed. The design is so that no two heads are the same, and that the prevailing colours reflect the orientation colours of the different floors. Common to all Generator hostels is the "funky G" logo – which the architects had lasered onto the tap heads in the bathrooms, and the oversized room numbers on the doors.

The ground floor offices re-use the old office lights from the previous use of the building—a good example of recycling.

085
IDEA-TOPS
艾特奖

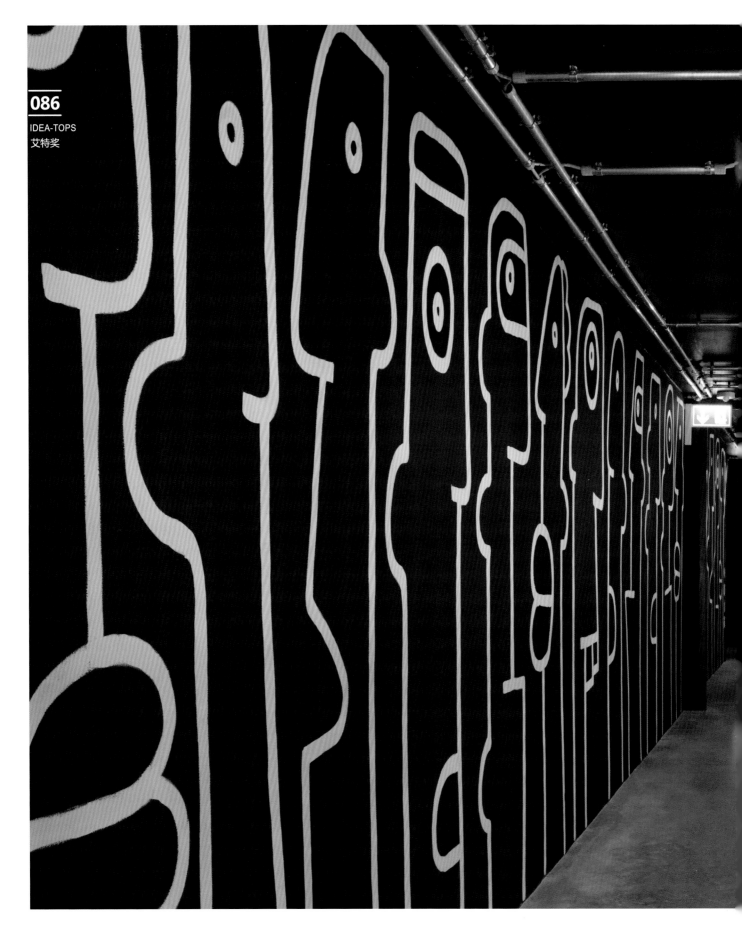

NOMINEE FOR BEST DESIGN AWARD OF HOTEL
最佳酒店设计提名奖

梁小雄（中国·深圳）

获奖项目/Winning Project

金陵天泉湖紫霞岭度假酒店
Jinling Resort

设计说明/ Design Illustration

金陵天泉湖紫霞岭度假酒店坐落于盱眙天泉湖岸边，湖光山色风景优美，落日时紫霞将照耀在整个酒店范围，故名紫霞岭酒店。酒店建筑依山临水，拥翠抱绿，整体造型为汉唐风格，酒店室内设计将新亚洲自然风格和天然美景融为一体，最为难得的是那份逐水而居的宁静致远。

金陵天泉湖紫霞岭度假酒店建筑面积2万平方米，设有大堂吧、全天候自助餐厅、中餐厅、宴会厅以及室内游泳池等功能，坐拥湖光山色、水景与露天休闲广场。酒店的设计是人性、健康、崇尚自然的，注重人与空间环境的交融。

接待大堂缀以暖色调原木板材，契合建筑本身的桁架元素进行空间分割，尽显精致高雅，从中可一窥酒店的新亚洲自然风格。墙面的石材取材于当地的火山石。经过重新切割后融入新的环境之中。

酒店接待大堂、大堂吧与室外休闲区水景联成一线，在这里可以欣赏到天泉湖的美景美色。在这里营造出内敛禅意的中式意境，身处其中可以感受到空间带给人的静谧和感官的伸展，整体自然的色调将眼睛和身心得到沉静与安宁。

碧水倒影晶莹闪亮，天泉湖美景绚丽夺目。沉醉于天泉湖紫霞岭度假酒店全天候自助餐厅，在彻底放松的同时重焕活力，在宁静平和之中享受味蕾刺激。

金陵天泉湖紫霞岭度假酒店的湖景房，犹如景观的有机组成部分，能够完全和谐地融入周围环境，每套客房都有宽阔的阳台和临湖景色的浴缸，现代自然风格的室内设计展现了独特的生活氛围。

日落紫霞，唯美的湖景线和精致的佳肴，当你漫步酒店内外，放松心情，感受无人打扰的休闲时光，时间都仿佛放慢了脚步，静静流淌，只在这处静谧的空间。

By the Tianquan lakeshore of Xu Yi city located the scenic Jinling Resort. In the evening, the sunset will beacon the whole range of the hotel in purple color. Sitting at the foot of mountain and embracing the greenery, it is a Tang Dynasty styled hotel. As for its interior, it blends Asian touch with natural elements to live up to tranquil environment and state of mind.

With 20,000 sq.m of area, the Jinling Resort offers lobby bar, 24-hour buffet, Chinese restaurant, ball room and indoor swimming pool etc, taking the fascinating lake and hills as the outdoor park. It features use-friendly, healthy and natural design.

In terms of the truss structure and holism idea, the hotel lobby takes on a warm tone by wood plate. Lava stone applied on the wall is produced at the local, then it is cut and sanded to fit into the new environment with an Asian look.

One could appreciate the picturesque Tianquan lake view in the hotel lobby, lounge and outdoor park, providing an implicit Zen prospect through a mild color to refresh one's body and mind.

The glinting Tianquan Lake, the deluxe 24-hour buffet of Jinling Resort, makes you linger over either in its peaceful nature or in its delicious cuisine.

The guest room that enjoys lake view as if a natural part of Jinling Resort, spacious balcony and bath tub facing the lake blend completely into the surroundings. Accordingly, the interior design represents a naturalistic life style.

When the evening falls, sunset lake view and refined cuisine are tailor-served for you. When roaming in or outside the hotel on leisure, the time seems slow down.

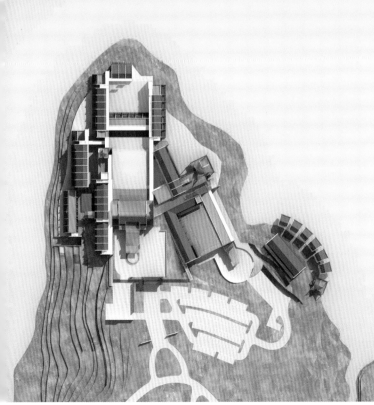

092
Best Design Award of Dining Space
最佳餐饮空间设计奖

艾特奖
最佳餐饮空间设计大奖
BEST DESIGN AWARD OF DINING SPACE

IDEA TOPS
INTERNATIONAL SPACE DESIGN AWARD

093
IDEA-TOPS
艾特奖

获奖者
BK ARCHITECTURE
（以色列）

获奖项目/Winning Project
Pastel餐厅酒吧 / Pastel Brasserie and Bar

094

IDEA-TOPS
艾特奖

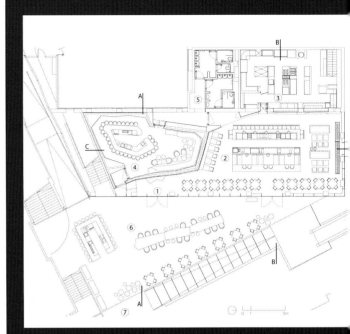

1. Entrance
2. Dining
3. Kitchen
4. Bar
5. Toilets
6. Outdoor Terrace
7. Sculpture Garden

获奖项目/Winning Project

Pastel餐厅酒吧
Pastel Brasserie and Bar

设计说明/ Design Illustration

Pastel餐厅是BK设计事务所的新作品，它位于特拉维夫的文化区，坐落在以色列特拉维夫艺术博物馆的翼端。本身Preston Scott Cohen设计的这所艺术博物馆就已经在传媒和社交媒体中引起了广泛反响。

BK公司则要在这个庞大的美术馆一楼的400平方米空间内，面朝着前方的雕塑花园和桉树丛的绝佳位置，建造一个与众不同的餐厅。

建成后的Pastel成了一个著名的城市餐厅，而不是一间博物馆餐厅。因为它不是按照博物馆而是响应城市的脉搏而对外开放的，俨然成为特拉维夫文化中心的一个独特的社交场所。

Pastel的入口与博物馆侧翼的公共广场连接起来，从外面穿过雕塑花园和公共广场即可到达Pastel。

Pastel的创作灵感来源于两个对立的世界：现在和未来。即表示在这极具几何形体建筑的羽翼一端的崭新世界和餐厅的世界，毕竟，餐厅作为传统的社交场所在西方文化中已经有200多年的历史了。

BK却将这两种世界以最直接的方式表现出来。富有张力的羽翼一端变成Pastel的室内空间，令这间餐厅看起来宛若乘坐时光机器着陆至此。

这两个世界皆疯狂地依附着这个DNA，恍然不知它们将是如此悦目的结合。"现在和未来"的世界是用延绵不绝的白色和充满活力的几何天花板构成。东南方的光线在灰色石制地板上跳跃，使天花呈现出柔和的质地与友好的氛围，这也是社交场所必备的感性条件。

另一方面来说，餐厅的世界又集合了古典的元素，如雅座、水晶吊灯、奢华的大理石台面、Thonet木质椅子和鱼骨木桌面等，它们呈现出波尔多葡萄酒的色调，唤起了对时间和空间的思考。

在Pastel的另一头是一间书房，被改造为葡萄酒藏书阁。可容纳一大群人在这片几何的云朵下享受隐秘的时光。

餐厅酒吧和室内同样可以眺望到露台和雕塑花园的景色。室外的椅子倾斜而置，连成一线。

食客和吧客共享着大理石桌面，模糊了他们来此的目的，甚至可以与长桌后面的其他人随意交谈。

Pastel拥有一个秘密场所。在这个包厢之内，一间"酒匣"很巧妙地逃脱了宾客的双眼。仿似一个局中之局，欢送出高酒精和高分贝的匣子。然而，这个迷你酒吧是有生命的，它遵循了两个时间规律，注定是城市夜行者的归属地。一个不起眼的门引领你进入迷你酒吧，超脱于餐厅的疆界，那是一个意想不到的放歌纵酒之地。

迷你酒吧是一个不大的会所。它的截面是餐厅全局最低的一侧，被精心设计成一个紧凑的360°酒吧。充斥着狂热的酒精气息，宽盈空间里的低位靠椅是为了让晚上来喝酒的人紧密地融入在一起。

波尔多葡萄酒的色调更加强了这种不绝的酒精气息。

全年开放的雕塑花园也推动Pastel成为一个标志性的社交胜地，露天酒吧和电镀桌台迎接肆意的活力和聚会，派对随时开始。

Pastel is Baranowitz Kronenberg's new brasserie on Tel Aviv's cultural block, located at the new wing of Tel Aviv Museum of Art. The design of the new wing by Preston Scott Cohen has already conquered the media and social networks, staring its daring architecture around the globe.

This is the context into which BK steps to develop a 400 meters square skeletal empty space placed at the ground level of the edifice facing the museum's sculpture garden and its grove of eucalyptus trees; an exclusive setting for a restaurant unlike any other in the city.

Yet Pastel is first and foremost a city restaurant not a museum restaurant. Privileged by a complete freedom from the opening hours of the museum Pastel responds to the pulses of the city and introduces a unique social venue for the heart of Tel Aviv's Cultural center.

Entrance to Pastel is via a public piazza flanking the museum compound and the visiting public is requested to step outside the museum and cross over to Pastel through the sculpture garden.

BK developed an indigenous narrative for Pastel inspired by two opposite worlds; the world of "new and next" represented by the hyper geometric architecture of the new wing and the world of The Brasserie - an almost two centuries old culinary and socializing bastion of the western world.

BK juxtaposes these two worlds in the most straight forward way. The dynamic envelop of the new wing is sucked into the space to become the interior envelop of Pastel while the Brasserie world seems to have landed out of a time machine right in the center of the space.

Each of the worlds jealously clings to its DNA, both unaware of the beautiful friendship which is bound to begin between them…..

The world of "new and next" is represented by a white and vigorous geometry clouding Pastel with endless shades of white. The south - east light bounces on the grey stone flooring and renders the ceiling with a soft appearance and a welcoming air, a must have mood for socializing.

The Brasserie world on the other hand is the embodiment of its classics; booth seating, crystal chandeliers, voluptuous marble tops, Thonet chairs which all come in Bordeaux color tone placed on top of a herring- bone wood platform, an island of sorts evoking different times and places.

At the very far end of Pastel a Study was developed shadowed only by a wine library where even large groups of people can share a sense of privacy under the cloudy geometry.

The dinning bar celebrates the views to the terrace and the sculpture garden as much as those of the interior scene. All guests face the view through a declining section of the seating arrangement.

The marble tops serve both the dinners and the bar team blurring the borders between them and motivating unexpected open-minded conversations with the team behind the counter.

Pastel holds a secret venue. Within its envelop, an "alcohol beat box" is concealed from the eyes of the guests. A box in a box, pumping high rates of alcohol and sound dbs. This Mini-bar has a life of its own which abides to a different biological clock. It is destined for the city's insiders, for those who live the long nights. An inconspicuous acoustic door leads into the mini-bar transcending the culinary world into the realm of the unexpected and of alcohol driven experience.

The mini-bar is designed as small scale cathedral. Its altar bears the lowest section of the space, programmed to build up high pressure and intensify the experience around the 360 degrees bar. A wild alcohol "cross" floats in mid air marking the reason for all to be there. Low seating is encouraged under the high section of the space with the intention of making the late nighters mix, mingle and connect.

The Bordeaux color of the space intensifies the alcoholic all-hours scene.

The sculpture garden terrace is a unique all-year venue which leverages Pastel as an iconic social hub for the free flow of energies, hanging out and the buzzing energies supported by an open skies bar and a 12 meters long galvanized steel table where partying is only imminent.

096
IDEA-TOPS
艾特奖

NOMINEE FOR BEST DESIGN AWARD OF DINING SPACE
最佳餐饮空间设计提名奖

周易（中国·台湾）

获奖项目/Winning Project

轻井泽锅物-高雄博爱店
Karuisawa Restaurant - Kaohsiung Bo'ai Branch

设计说明/ Design Illustration

从喧嚣尘世进入静谧时空的过程里，逐步获得适度转换、沉淀，沿着脚下温暖的实木栈道前行，眼前即是一番"独坐幽篁里，弹琴复长啸"的忘我意境，打造宛如长形托盘般的飘逸水景，池内缀以峥嵘景石、灯光喷泉与植栽，并以精致步道灯镶边，颇有怀石料理摆盘的风雅布局，一幅情味无限、姿态潇洒的现代水墨嫣然生成。

透过多层次榫接聚合的巨大木结构，诠释犹如殿堂般的藻井意象，使量体轻盈的浅色系，和环境主色调的浓重形成强烈对比，而天花板的木结构与中央卡座分界的手作金属格栅，两者呈现既冲突又和谐的90度结构视角，令人由衷感受到空间本质的浩瀚雄浑。过道旁一列以悬挂草料桶装，内置成堆珠圆玉润的红柿，堪称最富和风况味的季节果物，昭告店家开市大吉、来客事事如意的诚挚祝愿尽在不言中。

In the process of going from the hustle and bustle of this world into the quiet course of time and space, the mind is gradually converted and eased, walk along the warm wood plank road under the foot, and then a mood of selflessness like the verses "Seated alone by shadowy bamboos, I strum my lyre and laugh aloud" appears in front of the visitors, a tray-like elongated elegant water feature is built, the pool is decorated with towering landscape stone, fountains, plants and exquisite trail lights are used to trim the edge of the walkway, showing the Kaiseki arrangement like elegant layout, an infinitely touching, sweet and chic gesture modern ink and wash painting like view is beautifully created.

The huge wooden structure aggregated through multi-level mortise, interpreting palace-like caisson imagery. The light-colored system which makes the body feel lighten is in stark contrast to the strong primary color in the environment, the wooden structure in the ceiling and the hand-made metal grid boundaries in the central seats, showing a both conflict and harmony ninety degree structure perspective, making people sincerely feel the vastness and forceful nature of space. Fodder barrels are hanged over alongside the road, rounded red persimmons are put inside the barrels, they are the richest flavor season fruits and they are used to proclaim that the store is opening its door for business, visitors and all the best wishes are sincerely self-evident.

NOMINEE FOR BEST DESIGN AWARD OF DINING SPACE
最佳餐饮空间设计提名奖

古鲁奇建筑咨询(北京)有限公司

获奖项目/Winning Project

鹤一烧肉
Tsuruichi Yakiniku

设计说明/ Design Illustration

Tsuruichi Yakiniku，是一家高人气的日本烧肉店，位于日本大阪东部的鹤桥地区，以其独特的牛肉烧烤闻名当地。那里是旅日韩国人与当地日本人的混居地，因此也形成了独特的多元文化。"鹤一"用日语解读的话，即为"鹤桥第一"的意思，更是近百年前鹤桥地区的第一间烧肉店。本项目条形隔扇的使用说明，在日本建筑语汇中条形隔扇一直是我们所常见到的，大量密集的直线一定程度地体现日本审美的方式，我们在收集设计素材时发现了日本浮世绘上对于雨的有趣表现方式，《江户名胜百景》是歌川广重的最后一组版画作，其中最著名的一幅即是《大桥骤雨》。更因曾被后期印象派大师梵高所临摹而令世界瞩目。图中捕捉大自然的瞬间变化及旅人的突然反应，密密麻麻一条条的线条表现夏天的倾盆大雨，这种艺术表现方式在西洋及中国水墨画是看不到的。
设计理念抽象地还原了日本大阪鹤桥地区拥挤的街道建筑形象，餐厅的两个入口安排在"山墙"（中国建筑词汇）上，进入餐厅三角形的坡顶映入眼帘，我们刻意压低房檐，意图用强烈的三角形房顶来表达日式建筑与山林的关系。 在日本建筑中,如果入口开山墙上，就叫做"妻入式"；相反地,如果是从平顺的房檐下进入室内，这种方式就叫"平入式"。 房檐的高度是2.1米，即使作为住宅，这个高度也很低。作为餐饮公共空间，让客人穿过这么低矮的房檐进出包房，设计这个高度被甲方投诉我们毫无常识可言，这个高度伸手都能触到房顶了。房檐设计这么矮，意在弱化墙面，强调错落的三角形房顶，墙壁消失于房檐的阴影下，凸显了在都市建筑丛林中的小小山林。

Tsuruichi Yakiniku is a very popular Japanese Yakinikum restaurant famous for its wonderful beef barbecue. It is located in Tsuruichi area, east of Osaka, where many Korean families live among the local Japanese households, and therefore, it sees a unique and diverse culture there. Tsuruichi, in Japanese, means "the top one in Tsuruichi area", and actually, it is the first Yakinikum restaurant in this area in the recent hundred years.

We used strip-type partition board in this project, because it is always used in Japanese architecture. And in fact, the use of numerous close-grained stripes, to some extent, satisfied the aesthetics of the Japanese people, because we found an interesting way of drawing rains in Ukiyo-e (a famous Japanese painting collection) when we were finding materials and inspiration for the design: the drawer Utagawa Hiroshige used dense strips to paint the pouring rains in Summer in the picture named Sudden Rain Falling on the Bridge of his Hundred Beautiful Sceneries in Edo Era collection. The sudden change of weather and the instant reactions of pedestrians were captured by the drawer and showed in the pictures, making a great painting collection and even attracting Vincent Van Gogh's attention and therefore, the picture collection became well-known in the world. This way of painting rain is unique also because you can neither see it in the Western paintings nor in Chinese ink paintings.

In the design, the restaurant echoes the crowded streets and dense buildings: the two entrances are at the gable walls, and when you are inside the restaurant, the tri-angle slope crest immediately comes to your sight. Also, we intentionally lower the eaves in order to use the tri-angle roof to indicate the relationship between Japanese architecture and the hill of woods. In Japanese architecture circle, if the entrance is in the gable wall, it is called "tsuma iri", otherwise, if the entrance is under the eave, it's called "hira iri". The eave of this restaurant is only 2.1 meters, which is very low even for a residential building, let alone this is a restaurant, a public building. So we were claimed by the Developer that we have no common sense of architecture, and the customers are able to touch the roof as long as they stretch their arms. But we explained to them that we made the eave so low is because we want to make the walls hided by the eaves and their shades, in this way, the tri-angle roof can be stood out and therefore make the restaurant look like a small hill among the modern buildings in the city.

NOMINEE FOR BEST DESIGN AWARD OF DINING SPACE
最佳餐饮空间设计提名奖

Teresa Sapey（西班牙）

获奖项目/Winning Project

雷诺餐厅
RENAULT RESTAURANT

设计说明/ Design Illustration

室内的灯饰采用了雷诺-威赛帝轿车的同款装置。
黑色的墙壁、天花板和地板，使整个空间都像是一个被遗弃的夜间高速公路。地板的表面涂了一层循环再用的轮胎橡胶。
长长的黑色餐桌更衬托出夜晚视角的独特气质。引人注目的吊灯设计是取自真实的汽车大灯，墙壁两侧纵向延伸的红色与黄色灯条，模仿了驾车的速度，看似更具未来感。
餐桌上面摆放的大餐盘，则印上了雷诺方向盘的图案。

The installation is dedicated to the presentation of the new Velsatis model of the Renault car manufacturing company.

The entire space symbolizes a deserted highway driven at night, with completely black walls, ceiling and floors. The floor surface is of recycled tire rubber.

The perspective of night is made even more interesting by the long black table that stretches beyond the mirror. It is lit by a spectacular longitudinal lamp made out of real car headlights and red and yellow horizontal bands of neon that run along the walls. They perfectly simulate speed and the Futurists idea of movement.

The table is laid with great plates embossed with a Renault steering wheel.

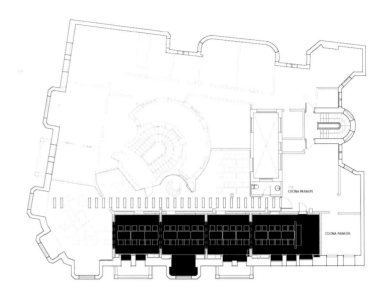

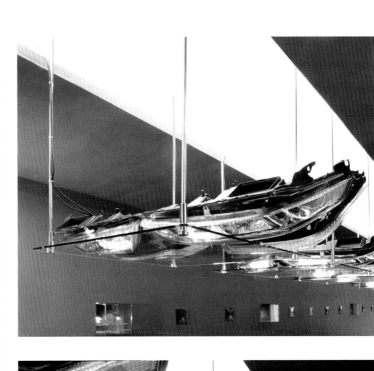

NOMINEE FOR BEST DESIGN AWARD OF DINING SPACE
最佳餐饮空间设计提名奖

陈德坚（中国·香港）

获奖项目/Winning Project

海玥餐厅
Café Bord de Mer

设计说明/Design Illustration

海玥餐厅位于香港大屿山的愉景湾，面积597平方米，是由陈德坚设计师设计的。由于餐厅位于滨海区内，所以希望为客人营造一个放松平静和雅致舒适的用餐环境。结合壮观美丽的海景，带给人们一个让心灵休憩的绿洲，以及享受一下轻松悠闲的生活方式。陈德坚设计师特意打造了大面积的玻璃窗让明媚的阳光进入，并且使餐厅显得更加开扬明亮，宽敞舒适，客人兼享壮丽海景和美味佳肴，顿感开怀赋有诗意。餐厅采用开放式的布局和互动式的厨房，并以原木为材料。天花板采用木百叶窗一直延伸到窗户，如此引人注目的设计，除了装饰空间外，还能分散直射的太阳光。同时，木百叶窗的思想也应用到空间的分区上，与天花板的设计互相辉映，让空间更加生动。艺术感十足的柱子是Café bord de mer的标签，其概念来源于传统的"漂流瓶"，人们将讯息放在瓶子里，扔到海里。设计师用写好讯息的信用金色丝带捆绑，成了柱子的型，夜晚它能呈现迷人的照明效果，信卷的文字也会显现在柱子上。海玥台更是一个融入自然的逍遥天地，诚邀来宾在此完全放松享乐。来宾可以一边享用特色鸡尾酒，一边欣赏红日落入海平线的浪漫时刻；又或一边在璀璨星空下浅尝小吃，一边细听海浪拍岸的旋律。桌子和地板采用自然材质进行布置和装饰，其中餐桌表面采用天然贝壳和硬质木材装饰。香港海玥餐厅给旅客提供豪华的度假体验和轻松的海岸氛围。

Located in Auberge Discovery Bay, Café Bord de Mer reflects an oasis of calm and a dignif lifestyle. Since the site lies within the coastal zone, the design objective is to create a relaxing a serene dining space that showcased the spectacular waterfront as a principal feature of its des and style.

To maximize the dramatic view of South China Sea, the restaurant is designed with the full-len windows that framed in all-natural components, inviting sunlight to stream in. Flooded with natu light this unusual setting imbues the dining experience with a seaside holiday romance.

The dining area is fresh and spacious with an open idea layout, interactive cooking stations a timber finishes. Part of the ceiling is topped with missive wood blinds, which has been extended the windows. It creates an attractive effect, splitting up and scattering the sun rays. Meanwh the wood blinds idea is applied to the decorative partitions which generate a vivid temperament.

The artistic feature columns are the signature of the restaurant. The concept comes up with tradition of "Message in a Bottle", that is, people put a message in a bottle, and tossed it into ocean. The feature columns appear to be rolls of letter papers tiding with a metal string, it a presents fascinating lighting effect which highlights the surfaces of the columns.

Beyond the eatery is a causal and family friendly lounge bar. The décor is cozy yet modern v soft carpet, wooden wine shelves and bookshelves, a marble-countertop bar and a light touc blue accent. Guests can relax in soft leather sofas with shell-themed cushions. The outd terrace is designed to bring diners closer to nature.

The space also features many nature-inspired elements including lightwood furnished tables a floors, natural shell displays on the dining tables and food menus made with hard wood covers.

This luxurious restaurant provides a mixture of holiday style and relaxed coastal atmosphere t is difficult to come by. It is indeed a wonderful escape from the reality.

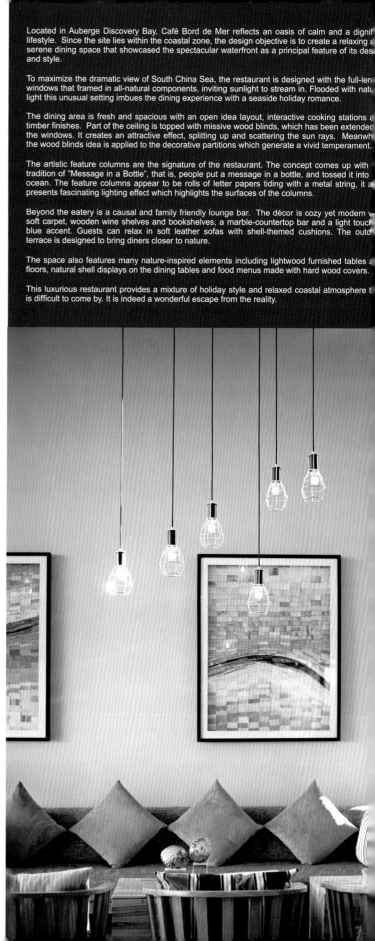

116
Best Design Award of Entertainment Space
最佳娱乐空间设计奖

文特奖
最佳娱乐空间设计大奖
BEST DESIGN AWARD OF ENTERTAINMENT SPACE

IDEA-TOPS
INTERNATIONAL SPACE DESIGN AWARD

IDEA-TOPS
艾特奖

获奖者
广州合壹设计有限公司

获奖项目/Winning Project
成都苏荷酒吧/ Chengdu Soho Bar

获奖项目/Winning Project

成都苏荷酒吧
Chengdu Soho Bar

设计说明/ Design Illustration

"铁器时代"主题设计，灵感源于英国维多利亚时期后工业时代经典设计风格，在酒与音乐构成的强烈场域中，以"静"、"冷"气质的"铁艺"作为主要装饰元素，铁艺的"冷"与氛围的"热"相互补充、调和，达到一种质感、温度的平衡。铁器时代设计充分尊重人与环境的关系，从多元感官出发，设计组合融合"铁艺"、"灯光"、"音乐"等多种元素，进入苏荷酒吧空间，仿佛进入一个交织过去与现在、自然与工业的独特美学空间，对于细节的极致打磨，无论从哪一个角度都能体会到一种令人放松的愉悦感。以铁器时代为代表的独具苏荷特色的娱乐空间设计，已经成为娱乐空间设计的经典案例，在业界一直被模仿，甚至引发了行业内的"苏荷效应"。

This Iron Times theme design is set in the Vitoria Age background. Alcohol and music are the body of this place. Cold as iron and Hot as atmosphere complete each other, compromising a balance of material and temperature. In regard of the relationship of people and environment, blending with elements such as light and music, you are entering a bar but also on board a time machine to an industrial era. You will be amazed by its cool details at every corner. Soho bar has already become an iconic design for the entertainment community and kept on creating her "Soho Effect".

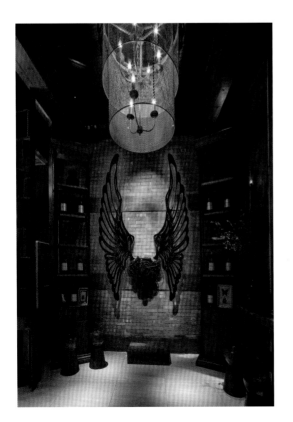

NOMINEE FOR BEST DESIGN AWARD OF ENTERTAINMENT SPACE
最佳娱乐空间设计提名奖

屈慧颖、冉旭、王清石（中国·重庆）

获奖项目/Winning Project

SOMEONE概念清吧
SOMEONE Concept Bar

设计说明/ Design Illustration

都市的夜空已离不开酒吧，它是每个城市对深夜不归的一种默许和亚文化的发生地。尤其是当我们进入E时代后，看惯了Windows的世界，过惯了机械式的虚拟生活，娱乐或自娱中的我们开始潮追起体验、视觉冲击、感官刺激……还有真实中的不真实感。酒吧更是成为人们重新解构平日真实生活的舞台。所以，在面对SOMEONE概念清吧的设计时，我们希望通过空间、视觉、音乐、互动、科技带来的感官冲击给城市的娱乐带来不一样的体验。对于这栋地处重庆江北繁华闹市的玻璃盒子建筑来说，任何奢华的材质、繁复的造型都显得多余，长达20多米、挑高近6米的炫酷空间，10多米长的巨幅投影在被继承了酒吧VI系统基因的几何形割裂开时，仿佛鼓点与音符幻化成了影像，恣意闪动奇幻的光芒。

在这个真实的空间里，光影交错更迭中更多透露出暂时的不真实，矛盾地对撞着，挑衅人们在这个"合理"的空间中进行情绪和压力的宣泄，人们在这里血脉偾张，难以自持。但与此同时，我们还应该明白，空间作为一个载体，它不应该只是一个视觉的容器，人和人在空间里的各种体验与感受才是主体，而体验不是自发是需要诱发的，我们意识到设计时需要找寻理想的体验媒介。所以，我们尝试在户外外摆区的围墙采用了智能光感玻璃，围墙可以随客人的需求变得透明或不透明，私密与窥探，就在随心的一瞬间。通过研究无线充电技术，们在设计时经过巧妙改良，在每一个吧台上客人可以把手机放置在台面上就直接充电，新奇却又实用的消费体验和利用科技带来的设计创新让我们和客人都兴奋不已……我们尝试不再孤立地去思考某一次设计或某一个产品，而希望通过各种手段和途径来创造一种综合的效应以增加消费体验。

当我们习惯了灰色的墙、灰色的天空、灰色的面孔和灰色的欢乐时，不禁想问问：这些连绵的灰色真能锁住我们对生活、对娱乐、对挥霍快乐、对释放压力的向往么？我们希望让SOMEONE概念清吧，聚合成城市中一种关于年轻的暗号，通过大胆的设计、想象、创新和探索，为人们创造出具有高辨识度和自己独特魅力的酒吧。

For city insiders, night is only a beginning. Being a part of the E era, we rely on the virtual life of the internet for entertainment and self amusement which blurring the border between reality and virtuality. And the bar is stage reflecting real life. Therefore, SOMEONE bar impacts nighters with her different space design, visual audio effects, music, state-of-the-art technology and engagement atmosphere. Located at downtown Chongqing, the crystal box like architecture that is 20 meters long and 6 meters high proves to be a sought-after place without question. With her more than 10 meters long screen at night, the beats and melody dance with the light, turning into a fascinating backdrop of the bar.

Light and images overlap just like the collision of people's emotion and tension bring unauthentic delusion to this existing space. Although it is known to all that the space is not only a vessel of luxuriance but also a spot to feel and touch, we try to find the perfect vehicle in the course of design. The outdoor intelligent optical glasses can be either transparent or opaque. To the visitors' great excitement, you can charge mobile phone on any table. We try not to focus on a single design or high-tech gadget, but to provide a new and comprehensive quality service and products to make consuming experience more enjoyable.

We have already got used to grey color scheme of the wall, sky, faces and pleasure, is that what we really want to spoil our feelings with? At any rate, SOMEONE bar is distinguished by her bold design and innovative experience expressing one of a kind attitude that only urban youngsters understand.

NOMINEE FOR BEST DESIGN AWARD OF ENTERTAINMENT SPACE
最佳娱乐空间设计提名奖

One Plus Partnership Limited（中国·香港）

获奖项目/Winning Project

香港CINE TIMES时代广场电影院
UA Cine Times Cinema, Times Square

设计说明/ Design Illustration

在21世纪科技发达的今天，拍电影全部采用高清数码技术，在这个全方位数码化的时代，旧式的胶卷早已淹没在历史的洪流当中。但胶卷曾在漫长的人类发展史里担当过举足轻重的角色，这个重要的历史任务谁也不能磨灭。因此这间戏院设计正以胶卷菲林作为主轴，唤醒那段被人遗忘已久的历史。
整间戏院以黑白作主色调，一大片白色以胶卷的姿态萦绕着四周，它时而铺张、时而沿着建筑物本身的形态，自由弯曲起舞，延续胶卷应有的灵活性，在戏院顺滑地穿梭。而幼细的黑色条纹随意地将白色的"胶卷"分隔开，令纯白色的背景衍生出不同大小的长方形，远看就如折叠起来的胶卷，起伏不一。
踏进影院，观众先看到一排黑色的"电影墙"，墙的左右两旁各挂有两个电视荧幕，让他们一睹最新的电影预告，再决定心仪的电影。当移步到购票大堂，修长而顺滑弯曲的柜台瞬时令来宾止住了脚步，那种弯曲的弧度跟墙上的"胶卷"竟然有种莫名的一致性，设计延伸了戏剧的张力之余，更添一份典雅。天花上纵横交错的特制黑色LED射灯，由8种不同长度的射灯组合而成，长度由1米至6米不等，每支射灯的方向跟角度也不一，在天花自由交织，感觉就如身在拍摄现场一样，观众突然不由自主地被牵引到电影之中，化身成片中的男女主角。由於不同的折射效果，营造出多个光与影的组合，观众追逐着灯的影子之时，亦不自觉地寻找着属于自己的影子。
由黑及灰色石组成的地板，其构图亦是胶卷概念的延伸，在地板交织出不同的图案。这次黑色成为"胶卷"的主色调，灰色就充当划分的角色，将一大片黑色划分成多个不规则的几何图案，有别于"胶卷"常见的长方形状，增添趣味。
顺着走廊一直走去，大堂的统一性犹在，好让观众在走进影院观赏影片前一刻，仍然继续沉醉在电影世界之中，为开场前做足"热身"。影厅将大堂及走廊所用"灯"的元素继续延伸，入面挂有长短不一的吊灯，它们射向不同方向，提升影厅的立体感，就算仍未有观众入座，都制造出高朋满座的热闹感觉。

The 21st century opens a gateway to the new era of modern technologies and innovations. With the widespread of 3D and even 4D high definition movies available on global market, the old method of filming has gone scarce, if not extinct. When thinking of the theme for this cinema, the designers want to trace back to the roots when filmmaking began. Back in the 19th century, photographers captured continual images and stored them on a single compact reel of film. This ancient object – roll films, was being symbolized all over the cinema, reminding the audience the long forgotten history behind the scene.

Splattered with black and white, designers use white as the main frame of the roll films, which is opposite to the usual black color. The white background engulf the entire space, motioning itself smoothly along the wall of the building. Its flexibility reflects the nature of roll films. Thin pieces of black stripes intercut the white surface, resembling the breaking and reunifying of roll films, forming rectangles of different sizes and shapes.

LCD movie screens playing the latest trailers greet the audience in the entrance, giving them news on the hottest movie trends. The edge of the table of the ticketing office bends smoothly along the ambient. The bending angles match those 'roll films' on the wall miraculously, which brings an extra dramatic vibe to the whole design.

Above head hangs special-designed LED spotlights of eight various lengths in black, ranging from 1-6m aiming at different directions. Light and shadows fulfil every corner; one might have mistaken it as the shooting spot of the newest film. Audience might fall into the illusion of fantasizing s/he is the lead actor/actress in a romantic comedy.
Down on the ground, the floor resolve to the original state of roll films, black being the principal color. Grey stripes cut the black space into different geometric shapes and sizes, contrary to the rectangles on the wall, giving the design a frolicsome note.

Inside the auditorium, spotlights of different lengths are directing all over the perimeter. They intersect along the wall, which creates an additional multidimensional atmosphere, hoping to provide the movie goers a warm and comfy viewing environment.

131
DEA-TOPS
艾特奖

NOMINEE FOR BEST DESIGN AWARD OF ENTERTAINMENT SPACE
最佳娱乐空间设计提名奖

陈武（中国·深圳）

获奖项目/Winning Project

LOVE 100 CLUB 大同
LOVE 100 CLUB Datong

设计说明/ Design Illustration

LOVE100酒吧是新冶组在原唐会酒吧基础上的全新升级。经营面积1200多平方米。新冶组设计以多年实践经验和锐意创新，结合当地本土文化，打破分解既存的陈旧空间形式、格局和模式。以最新潮的娱乐文化理念，演绎激情张扬的空间调性。时尚动感的节奏，融视听之享受，更好地迎合当下大同娱乐消费市场。

风格主张新旧融合、兼容并蓄，整体偏于新派电子风，同时不失中庸之道，与大同这座古城遥相呼应。大厅矩阵灯光应用演绎先锋概念的灯光艺术，圆形场布局带来的是包围式的怡然放松。设计师以曲线和非对称线条为最小单位，在设计细节中把玩非理性因素带来的反叛、刺激和调侃。花叶等自然意向在墙面、栏杆、窗棂和家具上的装饰应用，赋予无机的世界有机的情调。整个空间立体形式都与有条不紊、有节奏的曲线融为一体。

360度全方位立体化、激情热舞现场，将国际尖端高科技与R&B、Hip-Hop文化多重混合，在科技电子和节奏舞曲中的缝隙空间，寻找最深入的摇晃点位，不单单是华丽的未来外壳，这就是LOVE100的自我性格，不动声色的一种强烈。

LOVE100 bar is the upgraded version of Tanghui bar from Newera Design. Business area of over 1200 square meters, Newera borrows the local culture and deconstructs the existing spatial configuration. Investing the latest entertainment philosophy and passion for design, LOVE100 is catered to the current consumer market.

In favor of fusion the old and new, moderation and extravagance, it features a techno style as a contrast to this ancient city. Installed with state-of-the-art matrix lighting, the hall layouts like an arena, comfortable and relaxed. However, there are some rebellious details such as curves and asymmetric lines. Flower and leaf pattern applied on the wall, railings, window and installation give this insentient entertainment world natural vitality. The entire space gives out a melodious coherence and rhythm.

On 360-degree and three-dimensional dance floor, where the international cutting-edge technology blends with R&B and hip-hop culture, LOVE100 is not only showy in appearance but also a place that you will love 100%.

NOMINEE FOR BEST DESIGN AWARD OF ENTERTAINMENT SPACE
最佳娱乐空间设计提名奖

Antonio Di Oronzo（美国）

获奖项目/Winning Project
Aura夜店
Aura Nightclub

设计说明/ Design Illustration

Aura是位于纽约长岛东梅多地区集灯光和音响为一体的一间夜店及聚会场所，致力于打造豪华的室内装饰和提供最先进的灯光与非凡的音响效果。

这个场所被构想成是光和声音的流动和振动空间，它卷入了涟漪、潮落和潮起之中。光影鼓动拥簇，四壁暗涌，在瞬息之间都静止了。

层状的垂面逐渐出现在夜店中央，曲面此起彼伏，弧形的网状灯光照亮了其巧克力色的树脂主体和以深红色树脂材料勾勒的边缘。棚顶镀铬的光面嵌板和LED网穿插在一起，其反射的光线营造了极佳的视觉效果，带动了夜店的气氛。

这个空间有一个跃式的设计并设有3个酒吧区。在稍低的一层中有一个公共舞池，如果说头顶的设计是属于面上的，那么地板的设计要归为线上的了。胡桃木制的栏杆将宴会厅旁的酒吧和桌子连成一线，里面的卡位是随机布置的。

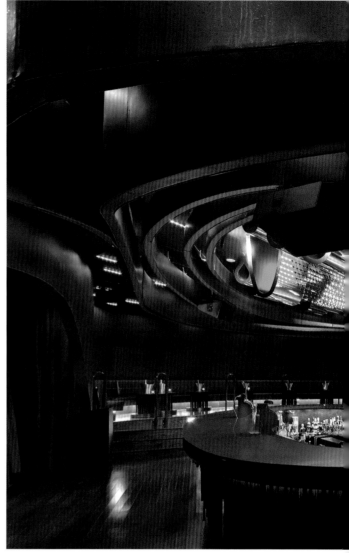

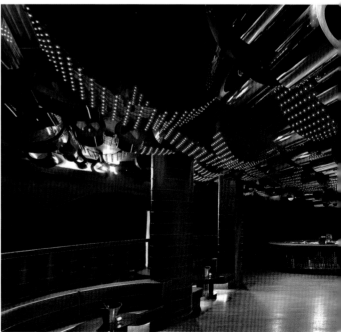

Aura Light and Sound Suites is a nightclub and event space in the East Meadow section of Long Island [NY], conceived to indulge the senses with plush interiors, state-of-the-art lighting, and crystalline sound.

The venue is conceived as a vibrating space that is as fluid and mobile as light and sound... interfering in ripples, waves, swells. Surfaces of light undulate and push, soft walls lift and reveal... all is ephemeral and light... momentary.

The space is layered with a series of vertical planes progressively rising as they approach the center of the room. As these reverberating surfaces rise and float, they reveal a slanted lighting grid that clarifies their materiality of chocolate brown vinyl and their edges outlined with a deep red vinyl. Beyond the innermost winding surface is an arrangement of undulating chromed panels and LED mesh screens playing and amplifying video signals and the vibe beneath.

The space is structured on two main levels and is fitted with three bars, one of which is free-standing at the lower level. If all the design systems from above are surface-like, all that is on the ground has a linear essence. Walnut rods line the bars and structure the tables by the banquettes, and are organized in a random arrangement.

140
Best Design Award of Exhibition Space
最佳展示空间设计奖

艾特奖
最佳展示空间设计大奖
BEST DESIGN AWARD OF EXHIBITION SPACE

IDEA-TOPS
INTERNATIONAL SPACE DESIGN AWARD

获奖者
尹杰、朱晓鸣
获奖项目/Winning Project
西溪壹号售展中心/ Xixi 1# Sales Center

获奖项目/Winning Project
西溪壹号售展中心
Xixi 1# Sales Center

设计说明/ Design Illustration

此案西溪壹号售展中心在西溪湿地原生自然美景的怀抱中，西溪壹号打造比肩江南会、西湖会等西湖畔会所的杭州首个世界级企业会所集群，汇集私密企业会所、高端商务、休闲娱乐等于一体，考量精英人士商务、社交、生活需求，形成西溪湿地之上的顶级商务群落。

通过精巧的设计，将景观向下渗透延伸，带南北通透的阳光露天庭院，首创"飞地"概念，让建筑漂浮在西溪之上，窗户外面创造性打造数百米空中水景，通过水面与绿植的视线控制，可不受干扰地一览西溪湿地公园全景，创造与西溪无边接壤，既开放又极度私密的禅意观景空间。

Embraced by the beautiful Xixi wetlands, Xixi 1# is a world-level company clubhouse by Xihu lakeside in Hangzhou. It is a confluence of private company club, premium end business and entertainment establishment, meeting the needs of business, socializing and living from elite personage.

This initiated "enclave" open courtyard is built afloat the Xixi wetlands, offering unobstructed prospect over the garden. It is an open-air space and connected with the garden seamlessly but its interior is private in contrast, full of Zen wisdom.

NOMINEE FOR BEST DESIGN AWARD OF EXHIBITION SPACE
最佳展示空间设计提名奖

秦岳明（中国·深圳）

获奖项目/Winning Project
南宁华润幸福里销售体验会所
Nanning China Resources Xingfuli Sales Center

设计说明/ Design Illustration

现代城市中心，繁华喧嚣之地，附和但不尽然。以空间之名，塑造心灵休息之所，以自然之意，构建城市绿洲。以"林"为主题，"简于形，而精于心，于形，而非于色"，结合现代艺术的表现形式，营造城市绿洲的氛围，引申出"远上寒山石径斜，白云深处有人家"的想象，让人沉浸其中。

时而如高耸矗立的大树，时而如蜿蜒交织的藤条，时而又如同罩上了层层叠叠的大网，光影交织，斑驳点点。配合黑色材质，营造强烈视觉冲击效果及神秘感，沉稳中带着新颖，高贵中透露着时尚，传达一种自然而然的心灵贵气，打造时尚与自然完美结合的高品质空间。

Is there no peace in the hustle and bustle city downtown? The answer is no. To build a shelter for the heart in terms of a house, forest is the concept and briefness is the style, conjuring up an oasis in the metropolis as if being in the middle of farthest mountain in the clouds.

As towering as big tree, as winding as cane and as cascading as huge net, the space has been overlaid with scene of lights and shadows. Black scheme brings about visual impact and sense of mystery mottled with innovation and elegance. It is a fusion of natural factors and fashionable modern space.

NOMINEE FOR BEST DESIGN AWARD OF EXHIBITION SPACE
最佳展示空间设计提名奖

刘锐韶（中国·佛山）

获奖项目/Winning Project

萝岗奥园广场销售中心
Luogang Aoyuan Square Sales Center

设计说明/ Design Illustration

面积：1500平方米
设计单位：广东锐美集思装饰设计有限公司

广州萝岗奥园广场是奥园集团拓展商业地产领域的重点项目，整体定位为具有国际化的品质和时尚尖端的体验式销售场所。本作品以金砖、腾龙、水纹为元素，通过更具创意的形式打造令人耳目一新的商业体验销售中心。

流线型设计赋予空间灵动感——流线型原是空气动力学名词，用来描述表面圆滑、线条流畅的物体外部形状。在最具商业气息的环境中产生的美国流线型风格，它的魅力在于它是一种走向未来的标志。独创的销售双动线设计，把"漏斗式"营销理念融入空间布局之中。接待大堂及销售大厅均采用中空设计，紧靠商业项目的主题，增强商业气氛的视觉体验。作品当中很多不规则的曲面造型采用了金属、人造石材加工而成，令人印象深刻。

Dimension: 1500sqm
Designed by: Guangdong Mei Ji Si Decoration Design Co.,Ltd

Luogang Aoyuan Square Sales Center is a key project of Aoyuan Property Group Limited focusing on a top level sales and service center. Based on the gold brick, flying dragon and water features, the commercial sales center takes on a refreshing look.

Inspired by streamlined design, its charm resides in the futuristic vision which comprises innovative marketing idea as well. The reception lobby and sales center with high ceiling enhance the product and business ambience, impressing the visitors by its flowing lines made of metal and artificial marble.

NOMINEE FOR BEST DESIGN AWARD OF EXHIBITION SPACE
最佳展示空间设计提名奖

刘国海（中国·广州）

获奖项目/Winning Project
广州时光里销售体验馆
Guangzhou Times Sales Center

设计说明/ Design Illustration

对于这个案子，我们的设计主张："传统"是依附在空间里呈现的场所精神,空间是骨架，"精神"是灵魂，而串连整个空间体验的"剪裁"——品质，将是现代生活和现代工艺结合后的结果。拒绝直白地将复古元素生硬并符号化地套进空间之中，一个"老茶馆"式的销售空间只能沉迷过往，却无法拥抱时代。传统和老去的痕迹应该是空间带出来的淡淡的情怀，不彰显、不具象。

查阅了很多广州的资料和老照片，意图在那些逐渐消逝或已经消失的记忆里寻找设计线索，从一堆资料中提炼出可以借鉴和运用的元素，并从西关老宅中提取空间的色彩和质感组合：用老榆木地板、非洲柚木索深色模仿旧西关民宅老化的木色痕迹；灰色的蓝贝鲁石材替代做岭南风格必用的灰砖，模糊那些比较具象和符号的套路元素；同时点缀大花白石材再现明清圈椅上镶嵌白色云石的细节。用全新的材料推演出一个有抽象传统记忆，但氛围和细节却具备现代情趣和品质的销售空间。

In this case, space is the body, concept is the soul and spatial experience is the quality. Our final work should be a fusion of modern life and up-to-date technology. We refuse to trap classic factors into design disregard of specific conditions, while an "old school" sales center is mere a boring duplicate and will be washed out by times. Therefore, traditional impression is better delivered by balanced approach. Referring to a vast of history and photos of Guangzhou, color scheme and materials are being extracted. Old elm and teak wood color like the Xiguan House, grey bricks as Lingnan architectural style and marble embellishment that they are the signs of our culture. Our memories are awakened by a series of abstract icons, but you can feel the ambience is temporary and you will be surprised to find out it is a sales center.

NOMINEE FOR BEST DESIGN AWARD OF EXHIBITION SPACE
最佳展示空间设计提名奖

KSL设计事务所

获奖项目/Winning Project

成都中洲中央城邦售楼中心
Chengdu Zhongzhou Central City Sales Center

设计说明/ Design Illustration

大手笔的挑空构筑出恢弘的空间尺度感，天花及墙面的欧式造型增加了视觉深度。璀璨的水晶吊灯缓缓洒落，与地面的巧妙设计交相辉映，形成一种荣耀光彩的联动。此售楼处设计以精粹奢华的材质镶嵌欧式新古典的精髓与高贵，以张弛有度的设计手法形塑高端诉求的品格与气度，尊贵与效益并存，古典凝粹与现代风尚交融，臻美空间绚华新生。

接待前厅
精雕细琢的天花墙面奠定了接待区的典雅高贵。吊顶与吊灯的设计给人以春华秋实的韵味遐想，天花与地面陈设及大理石拼花巧妙呼应，两侧少许金色、银色与一些反射型材质华丽调谐，整个空间韵律无穷。

模型区
8米高的挑空空间营造出磅礴气势，沙盘模型坐落于中心位置。华丽的水晶吊灯与窗帘让大理石铺设的宏伟空间显得尤为华美，多元素的糅合力求为来访者呈现一个独具舒适感和尊贵感的展示空间。

洽谈区
以简饰繁的欧式新古典陈设的尺度与节奏控制得精准到位。造型感极强的天花造型和华美吊灯的阵列拉伸了空间的比例，在灯光色彩的烘托下，营造出高贵的古典文化气息和强烈的视觉张力。

走道
走道凸显区域之间的融合与通透，细节的艺术感与精心打造的空间融为一体，每一处设计无不契合了高端人群的审美趣味和生活追求。

The massive void structure amplifies the airy space while European styled wall and ceiling increase its visual depth. Light pours down through the crystal chandelier reflecting its gloss on the floor. This sales center features elegance of neo classic Europe approached by fine design. Classic essence and contemporary fashion burst out new vitality.

Lobby
It is a refined reception space with crafted ceiling and wall. Golden ceiling and tree structured chandelier remind us of the harvest season. Glossy marble floor reflects with silver and gold deco composing a harmonious rhythm.

Showcase
Sand table placed at the center, the 8 meters high space takes us breath away the moment we walk in. Gorgeous chandelier and curtain underscore it with more elegance. Diversified indigents are dedicated to create a unique and comfortable exhibition space.

Meeting lounge
The European styled installation is simplified in accordance with the holistic amenities. Illuminated by the light fitting, it fulfills a desirable visual delight with its neo classical interior.

Aisle
The aisle works as a bridge connecting the other area in coherence with the concept. Every detail of this space narrates aesthetics and lifestyle of a higher level.

获奖者
利宾空间设计（深圳）有限公司

获奖项目/Winning Project
Pure33璞岸会所/ Pure33 Garden Clubhouse

获奖项目/Winning Project

Pure33璞岸会所
Pure33 Garden Clubhouse

设计说明/ Design Illustration

异形空间，一种颇具视觉冲击力的建筑形态和配合建筑表现形态的变化而形成不同的室内空间，是由利宾空间设计首席创意总监洪忠轩先生用独特的原创装饰设计手法使空间化异形为整形。

洪忠轩先生运用室内多种设计手法的灵活多变，把全新的设计潮流融入流线美感的室内异形空间里，使得现代人更舒适地享受那一丝属于自己的空间。异形室内空间包括室内平面布置的不规则形状、空间转角处出现的角落空间、立体上不规则的形状、天花出现的异形形态，几个因素相互结合所产生的奇形怪状，使空间达到显得不规则的基础上进行美感的体现设计，更能诠释整个空间因为不协调所产生的不一样的冲击力和遐想空间。

Alien space, a considerable visual impact of changes in the shape and form of buildings with architectural expression and the formation of different interior space, Living space- designed by chief creative director, Mr. Hong Zhongxuan creates original decorative design in a unique way to make space for the shaping of profiled.

Mr. Hong Zhongxuan uses interior design techniques in a variety of flexible, integrated into the new streamlined aesthetic design trends shaped interior space, making it modern and more comfortable to enjoy their own space hint. Indoor space includes irregular shaped layout of the room, there's space corner space, three-dimensional irregular shapes, the ceiling shaped morphology, several factors combined with each other grotesque generated, making the space seem irregular reach conducted on the basis of aesthetic design reflects better interpretation of the entire space because it is not the same impact and the resulting lack of coordination daydream space.

IDEA-TOPS 艾特奖

NOMINEE FOR BEST DESIGN AWARD OF CLUB
最佳会所设计提名奖

张莉宁（中国·台湾）

获奖项目/Winning Project

宁静山水
Peaceful Landscape

设计说明/ Design Illustration

"我期望建筑要能触动心灵，如诗一般。我们在其间体会了完整的时间感及存在感，一趟纯粹的灵魂与身体的旅行，是一种诗性化为空间经验的过程。"以内隐的姿态，冀望从历史文化底蕴中撷取灵感，以勾勒一座设计师心中的"逃城"，以呼应人们无法割舍都会生活却又希望抛却都市喧嚣的深层渴望。

此案以黄公望的富春山居图为发想，表达出宁静的文人山水。在建筑与空间设计上以素朴的语汇，静谧地植于土地之上，淡淡地融入这座喧嚣的城市中。设计师期望，建筑要能触动心灵，如诗一般。当建筑像是世上的"逃城"，在其间则可体会到完整的时间感及存在感。在这座"逃城"内，可自在漫游、放空自己，纯粹体验着空间与环境、自然和时间之间的感性对话。室内以传统合院之平面布局，左右对称，自大厅中轴线向后延伸，则来到后方的生活体验区。大厅左侧则以走道连接至样品屋，延伸出庭院的布局形式。最特殊之处是将接待柜台退于大厅主墙之后。大厅内，透光窗棂随时间洒进丰富的光影变化为丰富表情。如仪式般的走道，引领观者洗涤复杂纷扰的思绪。而双圆交错的大窗，刻意模糊了室内外的关系，也创造了属于东方空间精神的韵味与风采。

"I wish the building touch people's heart like a poem that we feel for time and our existence. The space produces a poetic atmosphere." Being inspired by the essence of history, the designer conceives an Escape Land that urbanites desire who cannot get rid of the metropolitan lifestyle.

Dwelling in the Fuchun Mountains is the work of Yuan-dynasty painter Huang Gongwang, it is also the intuition of the designer to touch people's soul. Leaving the sophisticated city and be simple, the downshifters could relax and empty their mind, just to experience where they exist and who they are.

The indoor layouts like a symmetric quadrangle house at the rear of which is the living area. On the left of the lobby is the sample room connected by a hallway, the most exceptional part is to set the reception desk behind the main wall in the lobby. Through the window lattice comes the light, painted with expressive tipsy. The hallway is as mystic as rituals leading visitors to clear their thoughts. The overlapped round window features a charm exclusively for eastern culture.

NOMINEE FOR BEST DESIGN AWARD OF CLUB
最佳会所设计提名奖

王砚晨、李向宁(中国·北京)

获奖项目/Winning Project

三苏祠博物馆景苏楼会所
Jingsulou Club

设计说明/ Design Illustration

景苏楼会所位于四川省眉山市的三苏祠博物馆内，其间绿意环绕，碧水迎风，优越的园林环境将建筑物收藏其中。原有建筑为二层砖木结构，曾作为博物馆的招待所使用，但由于多年闲置，庭院内杂木丛生，水道淤滞，给设计带来极大的机会及挑战，此次改造的目的，是将景苏楼打造为综合性的高端休闲空间，为博物馆提供新的服务功能。

设计工作围绕中国文人的庭院生活展开，苏东坡是宋代的文坛领袖，当其时，呼朋唤友，快意人生。景苏楼会所也寄期望成为今天中国上层精英人士的首选会聚之地。

室外空间的设计将传统的造园手法与当代的审美需求相结合。曲折的回廊将两处院落分隔开，但又形成空间上的连续性，庭院中的瀑布成为主景观，倾泻而下的水流形成动感的韵律和美妙的音阶，成为整个院落的中心，整个庭院充分体现中式园林"移步换景"的手法，每走一段路或转个弯都会有不同的视觉听觉体验，"随机因缘，构图成景"这也是中国式庭院生活的精髓。

室内空间的营造同样以庭院为中心，中国的文人是为庭院而生的，居于室内，窗成为内外景致连接的媒介。于是窗的材料，选用了近似传统窗纸的夹绢玻璃。透过格栅，院内的景致隐约可见，形成了梦幻的意境。室内的空间设计尊重中国古建筑的内空间结构，充分体现了中式古典建筑的结构及空间美感。

在室内材料及家具的使用上，注重选择有细腻质感的材料，如珠粒壁纸、丝质皮面、石材马赛克等，塑造出古典优雅的高贵室内空间。协调淡雅的色彩搭配，更是契合中国文人生活意境的品位需求。

Jinsulou Club is situated in Sansuci Museum, with green trees and blue water surrounded by superior garden environment. The original building was a two-storey stone and timber structure served as the guest house of the museum. Due to being idle for many years, the courtyard was dense with bushes and silted waterway, which brings great opportunities and challenges for the design. The objective of the renovation is to turn Jingsulou into a comprehensive and high-end leisure space to provide new service function for the museum.

The design is focusing on Chinese literati courtyard living. Su Shi was a leading literary of the Song Dynasty. At that time, he would often entertain his friends to have pleasure in life. Jingsulou is also expected to become a preferred gathering place for Chinese elites.

The outdoor space design combines traditional landscape techniques with contemporary aesthetic needs. Winding corridors separate the two courtyards but is formed with the continuity of the space. The waterfall in the courtyard becomes the main landscape, with water pouring down forming dynamic rhythm and beautiful sound. The whole courtyard fully reflects Chinese garden approach with each stretch of road or turns having different views and experience which is the essence of Chinese courtyard life.

The creation of interior space is also centered on the courtyard. Chinese literati were born for the garden. Living indoors, the window serves as a medium connecting the outside views. Thus, traditional window paper-like silk glass is chosen as the material for the window. Through window panel, the courtyard scene can be faintly seen, forming a dreamlike sight. The interior design respects Chinese classic interior space structure and fully reflects Chinese classical building structure and spatial beauty.

Delicately textured materials were selected for the use in the room and furniture such as beads wallpaper, silk like leather, stone mosaic which creates an elegant and noble classical interior space. Coordinated simple and elegant colors are quite in conformity with Chinese literati artistic taste of life needs.

NOMINEE FOR BEST DESIGN AWARD OF CLUB
最佳会所设计提名奖

深圳市派尚环境艺术设计有限公司

获奖项目/Winning Project

成都永立国际会所
Chengdu Yongli International Club

设计说明/ Design Illustration

宁静的诗意，采撷于自然

本项目位于天府之国成都，蜀地作为道教的重要起源地，其文化历来重视"无为"，其要义是："道"乃宇宙万物的根源，"道"是"无为"而自然的，智者应该而且必须体会天地自然的规律，顺其自然地把握自己，成就完整的人生。

因此我们希望在会所的室内设计中体现出"虽有人作，宛自天开"的和谐与平衡，保留自然的美感。木与石的纹理之美，以及玻璃的清透质感，在光影下呈现出繁复不规则的变化，透出典雅的氛围。也让人联想到高原海子澄净透亮的光影，山川河海变幻莫测的肌理。宁静的诗意，常常采撷于自然。因此室内主题雕塑、主题艺术装置、拼花纹理的灵感也源于大自然的启发与馈赠。在一楼入口大厅，大型木质装置艺术成为空间的视觉爆发点，模糊时空感和真实性，让整个空间有张有弛，充满节奏的变幻。

Serenity is extracted from the nature.

Chengdu is the Land of Abundance. As Sichuan is the origin of Taoism, she always makes much account of "inaction" which means Tao is the mother of universe and "inaction" is just spontaneous reaction. The wise should understand and stick to that philosophy in order to achieve the wholeness of life.

Consequently, we tend to exemplify the interior a reflection of the uncontrived nature. The natural texture of wood, stone and glass take on irregular pattern under the sunlight, as graceful as the shade and skin of the mountain and river just like the extraction of the nature. Accordingly, the sculpture, art installation and mosaic pattern are inspired by the nature. The huge wooden art display at entrance hall on first floor has led the visitors' visual delights to the zenith.

NOMINEE FOR BEST DESIGN AWARD OF CLUB
最佳会所设计提名奖

沈智立（中国·深圳）

获奖项目/Winning Project

南宁云星钱隆首府销售中心
Nanning Yunxing Qianlong Manor Sales Center

设计说明/ Design Illustration

建的"聿"，筑的"竹"，两者均有伸长、生长之意。从前，古人便把这寓意烙入建筑概念，而在本案的建筑外观，其简洁几何体的穿插、组合形成的体态便诠释了该点。在室内，为了延伸这份现代、简约以及生态的概念，我们"移接"了仿生的环境概念。我们认为的设计感并不是静止的，也不是每帧停止的感光菲林，而是"嗒嗒"转动的电影胶卷。它能与人们交流、互动、传达情感、唤起思绪，以其极致、简单、纯粹的设计美学，表达生活之中的精神信仰，以简约之态与自然相互交融。

本案云星钱隆首府销售会所项目，设计面积1390平方米，位于广西南宁市五象新区，毗邻南宁艺术博物馆，与AAAA级景区——青秀山森林公园隔江相望，具有丰富的自然景观资源。在空间处理上，我们优化了原有采光面积，带来大面积的玻璃幕墙、楼梯处高耸通透的幕墙采光井，为的是将更多的自然光线与景观引入室内中，使内外融会贯通。

在大厅主轴上，一朵迎露初放的代表南宁文化的朱槿花，中心垂吊的无数"花蕊"，似回报大地恩泽般洒下缕缕"生命之光"。整洁无瑕的线条、造型营造出行云流水般的自然韵律，合符其理的空间布局，亦能在行走间感受其柔美的动线。中央沙盘似被阳光拉伸的木饰投影，"迎风而弯"的金属造型板，写意的楼梯，或是看似随意摆放的球形绿植，充满张力的与墙体一体流动的服务总台，都在极具简约、现代感之外保留了空间原始形态之本。

无论从整体还是细节，从内里到外表，我们都希望能带来一股新锐大气之风，同时也希望竭尽所能拉近室内外的距离，或许，亦能拉近人、建筑与自然的距离。

Building, in its Chinese definition, holds a meaning of stretch and extension. The house consists of rectangular volumes and its interior represents a continuation of the existing modern yet ecological concept. The designer considers it not a motionless space but a restless movie films that mingles into the circumstance communicating with people.

Yunxing Qianlong Manor Sales Center covers an area of 1390 sq.m located at Wuxiang district of Nanning city in Guangxi province, adjacent to Nanning Art Museum and Qingxiu Hill Park the 4A scenic spot. Thanks to this excellent natural condition, a series of glass wall and courtyards by the stairs are schemed to enhance the indoor lighting, bridging the touch of outdoor.

Approaching the lobby, the pistil of hibiscus flower at the center which symbolizes the culture of Nanning is just like the sunshine radiating towards the earth. The beautiful lines not only conform the spatial composition but also lingering in people's heart. Flexible metal plate, informal staircase, arbitrary ball-like plant and concise reception desk all stay their natural form.

We want to introduce the context of the overall and be attentive to details, meanwhile great efforts have been put to narrow the distance of the building and the nature.

188
Best Design Award of Office Space
最佳办公空间设计奖

文特奖
最佳办公空间设计大奖
BEST DESIGN AWARD
OF OFFICE SPACE

INTERNATIONAL SPACE DESIGN AWARD

获奖者
LIKE ARCHITECTS
（葡萄牙）

获奖项目/Winning Project
Kinematix公司/ Kinematix

获奖项目/Winning Project

Kinematix公司
Kinematix

设计说明/ Design Illustration

Kinematix是一个IT公司总部的空间设计，该公司致力于保健和健身技术的发展。

这个项目的基本理念是灵活的、不间断的车库拉门设计，创造了多重的空间隔断，得以展示出多种布局的空间结构。

以水泥地板和纸浆质地的天花板为基调，连续的或断续的金属墙面给空间带来了不一样的元素。

三扇流动的落地车库门可以使空间完全地开放或分割开来。其中一些门能够同时滑动，让空间更具机动性。为了使自然光线照射到最里面地方，门上还设有窗子。

当人们在里面行走的时候，双排滑动金属门会给人以不同的色彩感受。每扇门的橙红相间的条纹图案也不尽相同，带来空间和色彩各式各样的搭配。

Kinematix is a spatial proposal developed for the headquarters of IT company Kinematix, specialized in developing technology for healthcare and fitness products.

The architectural intervention is based on a continuous flexible system of sliding garage doors that create multiple spatial partitions, able to (re)organize the space in many possible layouts.

Consisting mainly on a concrete floor and on a paper pulp projected ceiling, the space shows on the continuous – or fragmented – metallic wall its differentiating factor.

Three lines of running garage doors allowing to range from a big open-space to a fully divided room created three layers of total-height spatial dividers. Some of the doors can simultaneously slide or rotate to open, allowing a bigger flexibility of the space. Windows were also created in the garage doors, so that natural light could reach the more interior parts of the space.

Taking advantage from the double-sided folded metal sheet surface, it was created a kinetic chromatic experience that changes when the viewer moves on space. Different orange-to-red tape stripes were added on the side-oriented sections of the metallic surface generating different space perspectives and colorful reflections.

NOMINEE FOR BEST DESIGN AWARD OF OFFICE SPACE
最佳办公空间设计提名奖

南京刘波原创工作室

获奖项目/Winning Project

现代简约之——洁
Modern Simplicity–Purity

设计说明/Design Illustration

空间规划： 本项目为一装饰公司办公空间设计，在空间设计上动静分明，打破常规的大厅式接待方式，根据客户的多样化设计了5个不同风格的VIP接待包间。

This is an office space design of a decoration company. The hall breaks the conventional style and encompasses both dynamic and static sides of the space. There are five VIP guest rooms of different ambience in accordance with customer's requirements.

设计概念： 因为简单，才深悟生命之轻，因为简单，才洞悉心灵之静。一切的简单都蕴涵着淡泊宁静的真实。本案的设计以"洁"为设计主题，删繁就简，整体色彩定位中性色调，进门第一眼就让人感到整洁明亮，没有太多繁琐的装饰，几何造型沙发上方悬吊镂空吊灯，斑马装饰画旁静坐抽象小人雕塑，简单的物品构造，安静清新，仿佛时间已静止，空间动静分明，忙而不乱。每一个画面，每一处细节，简而不失其华，约而不显其涩。

We perceive the lightness of being because of the simplicity; we hear the voice of the heart because of the purity. Simplicity is all, and everything contains simplicity. It is also the spirit of this project. On entering the entrance, one will see a clean and bright space without much decoration. Pendant lamp that hung over the geometric sofa, abstract statue of a man sits beside a painting of zebra, they are simple and refreshing amenities. It seems the time has ceased, but every picture is rich in details.

NOMINEE FOR BEST DESIGN AWARD OF OFFICE SPACE
最佳办公空间设计提名奖

ALFONSO CUADRA KAYSER
（墨西哥）

获奖项目/Winning Project

上海CCG公司
CCG Creative Office

设计说明/ Design Illustration

这个项目是将一个现有的三层厂房改造成一个富有创意的现代办公室，把原来厂房中有特色的夹层楼面、挂屏利用起来并加入了中国传统的元素，最后呈现出一个吸引力十足、活泼的办公空间。顶楼加装的玻璃使整个楼层面积更加宽阔，也让建筑外部看起来更加具有冲击力。

主要任务
· 保留厂房的主要元素，实现阁楼改造。
· 加强一些区域的垂直流线感。
· 结合被动式通风及低排放暖通空调系统。
· 办公室包括各种功能性区域，如会议室、吧台、私人会议室、休息区和其他设施。
· 使用100%回收木材。

The brief for this project was to transform an existing three storey industrial building into modern office spaces with a contemporary and vibrant interior. The end product utilizes mezzanines, hanging panels, restored features of the original factory and traditional Chinese elements resulting in an attractive and playful array of office spaces. The addition of a new glazed fourth storey on the top level has increased the total floor area and given the building a more dynamic exterior.

Main Challenge-
·Maintaining the integrity of the factory elements, and fulfilling loft office needs
·Reinforcement and vertical circulation additions in some areas.
·Passive ventilation systems combined with low emission HVAC systems
·Combination of many programmatic requirements such as meeting rooms, a bar, private offices, areas for relaxation and other facilities within a larger functioning office building.
·100% reclaim wooded material

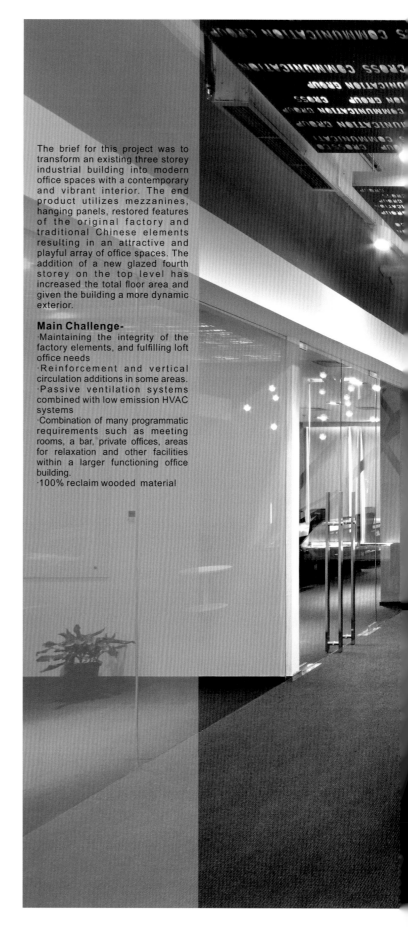

NOMINEE FOR BEST DESIGN AWARD OF OFFICE SPACE
最佳办公空间设计提名奖

上海鼎象装饰设计工程有限公司

获奖项目/Winning Project

鼎象建设集团办公楼设计
TRIANT DESIGN Office

设计说明/ Design Illustration

空间设计，根植于生活、存在于生活，是对平常生活独特的理解，是以细枝末节的形式渗入平日点滴之中。色彩的搭配、空间的转换穿插、家具的形态……都是设计的存在。如何向人们具象地诠释出设计与生活的关联，是这个项目空间思考的核心内容。

该项目位于上海徐汇区宜山路，是一个面积大约为2000平方米的办公空间。为了达成空间最大化的通透和明亮，设计师简化了材质的种类和形式：大面积白色墙面让空间非常明亮，木质的天然色彩则带来温暖安逸的氛围。来自希腊的爵士白大理石拥有优质纹理和光泽，在接待区吧台、过道墙面、地面都大面积的运用。

1.公司门厅，开阔明亮，线面简洁，时尚、活泼的现代设计风格。

宽敞的门厅，空间以纯净为主，立面、地面使用爵士白大理石相叠，灯光背景，立体光墙。吊顶天花大胆地使用不锈钢英文拼接，为空间带入强烈的前卫感，拼接字体正是鼎象的英文名称TRIANT DESIGN，呼应鼎象的独特个性，也同时解决了层高低、设备管道多的弊端。

透过用透明玻璃阻隔的VIP会议大厅，望向西墙上油彩手绘"TRIANT DESIGN"与之遥相呼应，再透过设计助理工作区域，向东看每一根承重柱上手绘的白描勾线画，又与西墙上的油彩手绘背景图互通连理，整个迎宾大厅四维空间，创意你中有我、我中有你，层层推进、融会贯通。

2.VIP会议室

步入上海鼎象国际设计公司的大门，迎接台的右手边就是VIP会议室，在这个部分，除了基础的采光外，为使空间更敞亮，运用地面高差和钢化玻璃隔断的形式围合出VIP会议室。玻璃隔断不但保持了两个空间的独立，同时也加大了VIP会议室与门厅的灯光互补。

两根承重柱与会议室的大门用老榆木本色饰面板装饰，榆木的自然纹路与爵士白大理石天然花纹在此不期而遇，和谐的自然肌理演绎着别样的情趣。老榆木饰面板与爵士白大理石在这个空间中被广泛应用，老榆木的年轮向人们诉说着它成长的经历，设计师大胆地将老材料新作；使老榆木不经任何雕饰和喷涂上色，本色地与爵士白大理石相处，这一现代设计手法被赋予太多的含义，自始至终不变的是纯净的质感，纹理独特，更有特殊的山水纹路，有着良好的装饰性能，在纹路走势、纹理的质感上，深藏着史前文明的痕迹，可以说老榆木、爵士白和玻璃牵手再一次将"新时尚"诠释了一遍。

设计师致力于打造一种轻松舒适明亮的氛围和环境，并将吊顶设计成一大片自然光的效果。发光膜与地灯交相辉映，会议室南墙上悬挂的120寸硬质液晶投影屏背景光也为会议室增添了不少光彩，起到了锦上添花的作用。

3.VIP接待室，尊重客户体验，粗犷中不失细腻，时尚中尽显典雅。

如果说大厅的设计属于大器典雅，那么接待室就应该是软玉娇香，也是整体风格的延续。一共4个接待室，空间面积相宜，现代设计风格不同，不论是几何块面的皮艺沙发，柔软细致的长毛地毯，还是经典现代的造型座椅，抑或是沙发背后错落有致的挂画，看似无序却有章可循，看似粗犷却十分细腻，温馨舒适中享受着现代典雅。

4.总经理办公室，格局简单灵活，氛围轻松，空间光线充足。

设计师遵循了简约的法则。在空间上最引人注目的恐怕就是右边洽谈区，天花、墙体到地面用老榆木地板铺设而成，这个设计是一举两得的，它即可以增加空间的层次变化，也可以突显一种现代奢华的美感。

宽大的办公桌背后，巨幅连绵起伏的群山油画，看似一幅油画实际上是一扇电动移门，移门缓缓移开，一个充满着文化底蕴的茶室呈现在眼前，在一个不足10平方米的茶室里有瀑布、鱼塘，还有日式的枯山微景观。正是由于其室外的河流，使得室内空间与周围的环境融为一体。席榻而坐望窗外河水荡漾、波光粼粼，尽思绪放松，品一口香茶，解一天劳乏……

5.设计办公区

从平面图上可以清晰地看到，整个空间被划分为两个主要区域：设计区域和工程区域，并有电动移门相连。设计区位于整个空间的左边。在这片空间中，四周用玻璃和老榆木隔出主要办公空间，中间安排了4张8米长的办公桌，几乎所有的设计员工都在这里办公，这样的方式强调了协作的重要。

办公整体使用白色。正因为白色有着微妙的包容性，它是颜色的合成，同时又是颜色缺失，因此避开颜色，才能唤醒空间最本源的物质性。

Space design derives from life and exists in life. It is the understanding of life and the details. It lies in the color scheme, space transition and furnishings. However, the core of this project is how to interpret the connection between design and life in a concrete way.

Located at Yishan Road, Xuhui District of Shanghai, it is a 2000 square meters office space. For the purpose to maximum the use of the natural light, the designer simplifies the form of the materials. The extensive white walls have lightened the whole space greatly and wooden color brought warm ambience. White exceptional marble imported from the Greece is planted massively at bar reception, corridor walls and floors.

I. Spacious hallway, a refined, vigorous and modern style
Pure color makes the hallway more spacious, the overlapped white marble enables the façade and floor a multidimensional background. The stainless-steeled TRIANT DESIGN letters which embody the spirit of the company on the ceiling bring avant-garde look to the space. Through the glass VIP meeting room, a hand painted TRIANT DESIGN on the west wall is well joined with the ceiling. The creativity is in everywhere, such as the graffiti on the king pillar in design assistant area and west wall.

II. VIP meeting room

On entering the TRIANT DESIGN office, the VIP meeting room is on the right hand of the reception desk. To enclose the VIP room by glass not only makes it a private space but also enhance the lighting effect by joining the hallway.

The two king pillars and door of the meeting room are decked with old elm over its original color together with the white marble illustrate a story of the nature. Old elm endowed with new life in this space creates a great but harmony contrast in color and texture to the white marble. To put in other word, that is the old joins the new. The designer is dedicated to forge relaxed and bright surrounding, taking the ceiling into natural lighting by the reflection effect. A 120-inch projection screen on the south wall heightens the appeal of the meeting room.

III. VIP guest room, value the experience of the customer, rough but elegant

The design of the grand hall can be summarized as elegant whilst the guest rooms delicately extend that style. There are 4 guest rooms in total. No matter how the geometric leathered sofa, soft carpets, characteristic modern chairs and murals distinguish from each one, guest could enjoy the coziness in the modern atmosphere.

IV. General manager's room, simple and flexible

The design follows the rule of simplicity. The most attraction is old elm applied in the business area on right side, the ceiling, wall and floor, adding diversified levels to the space. Behind the wide desk, there is an oil painting landscape of mountains where actually a tea lounge hidden behind the sliding door. In the space that less in 10 square meters, you can find a mini landscape of waterfall, fish pond and Japanese dried mountains and rivers. Sitting and looking outside of the window while having good tea, you will refresh yourself after one day's of work.

V. Design area

The space is divided into two main areas: design and engineering area connected with an auto sliding door. The design area on the left side is separated by glass walls and old elm. There are four pieces of ten-meter desks placing at the center. To emphasize the importance of team work, almost all the designers work in this area.

White is the theme color of the office. For the reason that white composes all the other color while it is absent of all the color as well, in other words, it shares the inclusiveness quality as the universe.

207
IDEA-TOPS
艾特奖

NOMINEE FOR BEST DESIGN AWARD OF OFFICE SPACE
最佳办公空间设计提名奖

何宗宪（香港）

获奖项目/Winning Project
Speedmark公司
Speedmark

设计说明/ Design Illustration

本案Speedmark是一所国际性物流公司，本案设计重点并非为企业本身打造形象而设，反而是希望利用工作环境促进员工与企业之间的关系。设计师利用物流行业本身的独特性质和运作程序，提出了一个适合物流的办公室作息生态。设计概念以connection (联系)为出发点，以连接身处于办公室处理文书工作的员工和在办公室外的船务海运码头以及机场空运同事，从而提高员工对公司的归属感。

办公动线
为强调"联系"作为设计方案的重点，项目在主入口的接待处用上货柜的造型(container)鲜明地带出公司背景性质，简洁的场景手法作为序幕，加上清晰的指示，带领着两个部门的人员一方指向"Air"（空运），另一方指向着"Ocean"（海运），同时拉近在办公室以外同时工作中的员工距离。进入员工办公的主要通道模仿升降起落的机场跑道，希望每位员工一踏入工作范围，仿佛是同时跟世界各地海运、空运的同事拥有共同的跑道。

办公空间
有别于一般趋向于高科技、无纸化发展方向，物流业仍会使用到大量的文书纸张，故此在室内设计一个达20米的长方形储物空间作为日常收纳之用，这不仅呼应着各个人员与货柜不可分割的亲密关系，同时消除了个别员工位置上储物的冗赘处理，使空间更为干净利落，而同事之间进行归档整合，都需要通过这个独立空间来做进一步处理，有别于昔日独自在自己的座位内工作，人与人之间缺乏交流。而这样的工作手法，如同在外派递同事般，向顾客派递包裹动作的缩影。配合着整个空间的清晰标志，更达致物流业上要求的速度标准。

休闲空间
物流业是一个讲求速度与准绳度的行业，为舒缓员工的压力，设计时在办公室特别规划出一个咖啡厅的空间，以运用warehouse(货仓)的概念主题作为切入点，这里不单纯提供员工休息的地方，同时亦是员工与员工、同事与上司之间一个交流的场所。这里以一个宽敞的空间，营造出一种可以从压迫的工作速度中释放出来的环境。每位员工在自己的位置里感受到与外地同事同样的氛围，透过不同场景铺排可以增加整个团队对企业的认同感，以及对公司品牌的进一步提升。

会客空间
在设计上把会议室当中的地台稍为提升，使整个感觉模拟为漂浮在海洋之中，让海洋的概念延续。天花部分则用上"天空"作为图案的拉幕天花，把整个的会议区天花的高度感觉提升，整体设计中选用了代表企业本身的蓝色，设计师亦透过不同的空间及饰面，进而以层递的手法处理空间之中的蓝色，分别为代表"海洋"、"天空"和"企业"。
整个设计方案以重视人员交流为出发，设计师以情景融合的手法，达致上述的要求，同时进一步营造出企业根本理念的氛围。不单展示出企业的中心理念，并拉近员工之间的距离，正如物流业本身，不只是货物之间单纯的收讫关系，而是一种人与人之间的联系。

Speekmark is an international freight forwarding company with representative offices in all major cities around the world. In designing its headquarter in Hong Kong, our primary aim is to emphasize the keyword "Connection" which forms the backbone of the entire logistic industry. Rather than superimposing a corporate image onto the space, the design concept is all about connecting people - bridging the office support staff and field worker staff at the ports and airports. We drew upon inspirations and elements from the industry itself and create a workplace that can facilitate staff's daily operation, and thereby nurturing among them a good sense of belonging to the company.

Elements pertaining to the logistic industry such as container cargo, runway, warehouse are displayed right up front at the reception, and are used throughout the entire office space. Upon arrival, the staff and visitors are greeted by a reception counter resembling a real floating container in blue, which is the corporate colour of Speekmark. Printed on the corrugated counter panel are two big words "Air" and "Ocean", which serve as signage to direct people to the respective departments of the company. As one proceed to the general office area, a long, narrow route mimicking a runway connects the individual working docks together. The underlying implication is obvious: it's a runway that is shared by both the office and field staff in different parts of the world.

An open plan concept is adopted in the general office to encourage communication and collaboration. To maintain the fluidity of the space, team supervisor's offices are defined by small level changes as platforms instead of visual barriers in order to establish the proper staff hierarchy. Considering the logistic company's huge storage need for paper documents, a 20 metre long storage room in the shape of a semi-open container is created and a creative backdrop to the whole workspace is formed. This centralized storage solution also encourages the interaction between staff who used to be confined to a cramped cubicle all day. Most importantly, it replaces the traditional filing cabinets scattered across the desks, and thus helping to keep the office organized, clean and clutter free at all time.

Continuing on the logistic design theme, the staff cafeteria takes on the setting of a warehouse in a sleek, modern expression. "Carton boxes" in different beige hues stack up as walls, cabinets and countertop. The soothing neutral colours, causal seating arrangement coupled with an unobstructed city view conjure up a relaxing environment to encourage more chance conversations to occur between staff from different departments.

The whole general office has a strong industrial feel to it, thanks to the use of untreated, utilitarian materials such as vinyl concrete flooring, corrugated metal sheets, raw wood and simple paints. On the other side of the office where conference rooms are located, a contrasting palette of clear glass, laminate mirrors, stone tiles and corona form a more refined, sophisticated corporate impression. This is where the clients come to visit, and the design thus befits its function to impress the clients with a strong corporate image for Speekmark.

This area is filled with warm natural light, and the use of reflective surface throughout further enhances this airy feeling. The conference room sits on a raised platform as if it is floating in the ocean. The ceiling is literally an LED light box printed with a blue sky image which helps to extend the ceiling height visually. Outside the room is a waiting lounge imitating a real aircraft - the bench seats are slightly raised, and the wall is carved with a row of airplane windows where the company's slogans are etched. Here, the colour of blue is used liberally as walls, ceilings and furniture, and it can be interpreted as the sky, ocean, or Speekmark. The office itself is an extension of the company's professional image and speaks volume about its logistic expertise and dedication to connecting people and the world.

212
Best Design Award of Show Flat
最佳样板房设计奖

文特奖
最佳样板房设计大奖
BEST DESIGN AWARD OF SHOW FLAT

IDEA-TOPS
INTERNATIONAL SPACE DESIGN AWARD

获奖者
大观室内设计工程有限公司（中国·台湾）

获奖项目/Winning Project
海洋都心3/ Ocean Heart 3

获奖项目/Winning Project

海洋都心3
Ocean Heart 3

设计说明/ Design Illustration

本案位于新北市淡水区的临海样品屋，业主期望打造此案为海滨度假居所，设计概念以此建案名称发想，将海洋相关的意象融入立面设计及材质选用之中。于空间规划上，则以休闲的接待会所为考量，将全室大半划分为公共区域使用，将客厅、餐厅、书房相互串联，自入口进来，即可感受到其大器的尺度，使亲朋好友于此交谊互动更为自在。立面设计则以圆弧线板等距分割的语汇，传达出船坞夹板规律的线条，借此修饰厨房及小孩房入口暗门，搭配典雅壁纸，以重复的分割与材质，强化空间延伸放大的感受。壁面与天花选用桧木染色木皮，以仿旧刷色的纹理，表达海沙质感的意象。主卧床头主墙特别挑选仿马毛质感的进口壁纸，配合床头上方的吊灯，赋予起居空间有如饭店的精致质感。小孩房则融入海军蓝、白条纹的意象，给予小孩房休闲兼顾活泼的氛围，并整合收纳及卧榻等功能。

Lied in Danshui district, Xinbei city of Taipei, it is the perfect sitting for a beach holiday. The designer blends the signs that symbolizing the ocean into the design and materials. To highlight its interactive function with family and friends, half of the space is planned as public space, while sitting room, dining room and library are connected in a consistent bearing. Signs of ocean are everywhere: lines of the space is as meticulous as the deck of a boat extending our visual senses, texture and color of wall and floor is as rough as the sand, navy blue and white stripe fabrics in children's room. To adorn master bedroom as refined as a hotel room, wall above the headboard is covered with imported faux horse hair textured wallpaper.

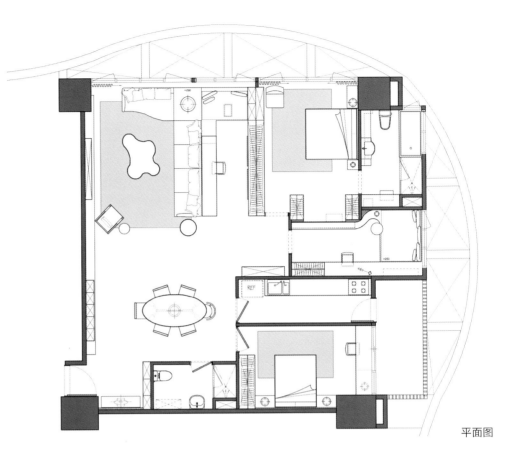

平面图

219
IDEA-TOPS
艾特奖

NOMINEE FOR BEST DESIGN AWARD OF SHOW FLAT
最佳样板房设计提名奖

韩松（中国·深圳）

获奖项目/Winning Project

CHOICE 选择
CHOICE

设计说明/ Design Illustration

深圳的生活
有太多的现实，
有太多的残酷，
有太多无休止的奔跑、追逐，
有太多的欲望魔鬼……
我们也许
无法选择财富，
无法选择成功，
也许无法选择喧嚣与否……
但是
我们可以选择自由，
选择随心而动的生活……

Life in Shenzhen is of too much reality and too much brutality,
Endless pursuit and restless desire,
Although we have no choice for wealth or triumph, or leaving the madding crowd,
We still have choice for freedom and follow our hearts.

NOMINEE FOR BEST DESIGN AWARD OF SHOW FLAT
最佳样板房设计提名奖

张鑫磊（中国·深圳）

获奖项目/Winning Project

态度人生(成都-御景峰度)
Attitude

设计说明/ Design Illustration

本案为楼盘系列小户型样板间的建筑面积45平方米户型，层高为5.5米，可以扩展为两房两厅的复式组合空间，面向群体为单身、或者新婚的年轻人的首次刚需置业。
空间运用深色木质与灰白色为主基调，既能满足空间现代、时尚的感觉，又有效地避免了纯黑白二色带来的空间冷硬的感觉。空间布局上采用了将大件家具固定在硬装的空间结构中，利用单件家具的设计感提升空间尺度的方法，避免了空间中堆积过多家具导致的拥挤与压抑，虽然只有45平方米的建筑面积，实用面积更少，但是整体的空间尺度感不输于常规三房的空间感觉。

This project is about a small show flat of 45 sq.m and 5.5 meters tall, a duplex space targeted for the first property of urban single bachelor or newlyweds.
Taking dark wood and grey as the color scheme instead of the cold black and white tone, it meets perfectly the needs of a modern and smart owner. Big furniture is built in the space while small designed pieces help introduce spaciousness to this small apartment, also to avoid crowding and repressing the space. Thereby, it is no smaller than a three-room's flat.

NOMINEE FOR BEST DESIGN AWARD OF SHOW FLAT

最佳样板房设计提名奖

柏舍设计(柏舍励创专属机构)

获奖项目/Winning Project

成都中德英伦联邦5#楼3302户型
Chengdu Zhongde 5#

设计说明/ Design Illustration

本案设计为顶楼三层复式的结构布局,以社交面广、注重生活品质的时尚人群为目标客户,融入国际化的视野,打造高品质的生活环境。
整体空间以深色柞木木饰面搭配浅色银狐木木饰面,再以白色钢琴漆做背景,大量的动态特征以深色调出现,在流畅的空间中蕴含沉稳,让居住者置身于动静皆宜的世界;顶层卧室是整个空间的亮点,立面的木格栅将影子投射到室内的空间,明亮而温和,微煦和风在室外将室内的影子吹动得婀娜多姿。通透的空间串连关系,玻璃天窗,简约悠闲的迷人气息在空间的几处亮蓝色中蔓延,演绎充满生命力的韵味。

项目地点:四川·成都
项目面积:约470平方米
主要材质:银狐木木饰面、柞木木饰面、白色钢琴漆等。

It is a three-storey apartment on the top floor, featuring sociable hipsters that enjoy international and quality lifestyle.

The whole space blends dark oak with light wood finish. Practice of the dark color streams amidst the air at fair pace unobstructed, leaving the white piano lacquered background stand still. Bedroom on the top floor lights the apartment through the window where introduces warmth and breeze in, communicating with the external world. Shadow of the railings scattered around, high blue color thrives in charm and vitality.

Details:
Location: Chengdu, Sichuan
Area: 130m²
Main materials: Yinhu wood finish, oak finish, white piano lacquer

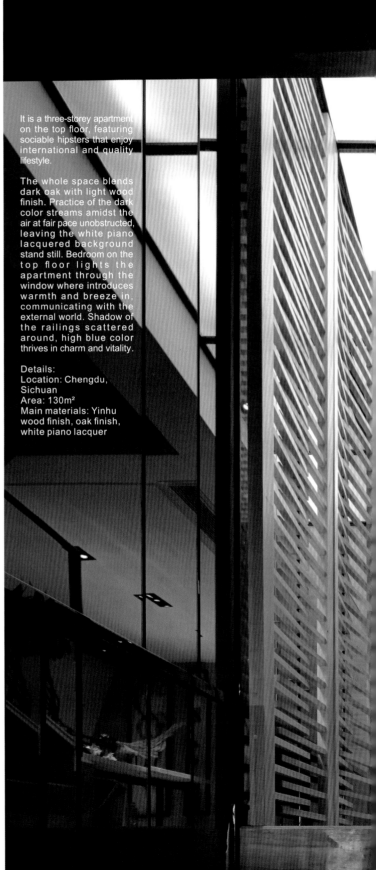

NOMINEE FOR BEST DESIGN AWARD OF SHOW FLAT
最佳样板房设计提名奖

Federico Delrosso（意大利）

获奖项目/Winning Project

摩纳哥公寓
APT AC MONACO

设计说明/ Design Illustration

高飞的房屋

最大化使用空间与突出一家人的日间活动区域是本案的出发点——这来自一个意大利家庭的要求，他们在摩纳哥的一间14层高、200平方米大的公寓里居住和办公，建筑从1960年代就已存在了。

设计师按照惯常的做法，通过连续的容器系统将整个区域和墙壁进行解构，采用与房间等高的房门。意欲在原有的家中打造出"新家"的效果，一系列房门缝隙间的照明营造了房屋整体的照明效果。

内层结构躲在承重墙之中，过道内窄细的玻璃的安置让人可以一探房间的深度，也让周围的环境得到延续。所有的构思都围绕着和谐统一。白色的基调，令人联想到美术馆，可将重要的设计或艺术品纳入其中。作为日间活动地带的延伸，露台变成了"高飞"的优雅海上客厅，绝妙的海景一览无余。

A high-flying house

Make the most use of the space and emphasis the daytime area where they spend time together: this was the request made by the Italian family living and working in the Principate of Monaco for an apartment of around 200 square meters on the fourteenth storey of a building dating from the 1960s.

Following his customary sartorial inclinations, the architect sliced up the total area via a continuous system of containers that blend into the walls, and which also integrate full-height with the doorways accessing rooms and services. Underscoring an operation that has created a "new home" in the old one, the deliberately monolithic tone applied to the place has been lightened by a series of illuminated and transparent gaps.

The inner shell stands shy of the support walls, and at intervals is interrupted by narrow gashes of glass that afford glimpses into the depths and lend continuity to the ambience. Everything is conceived around an idea of general harmony. A white box, similar to an art gallery, devised to harbour important pieces of design and art. An extension of the day zone, the terrace is an elegant "high-flying" nautical sitting-room given the height of the building and the admirable view of the seashore.

236

Best Design Award of Villa

最佳别墅豪宅设计奖

文特奖
最佳别墅豪宅设计大奖
BEST DESIGN AWARD OF VILLA

IDEA TOPS

INTERNATIONAL SPACE DESIGN AWARD

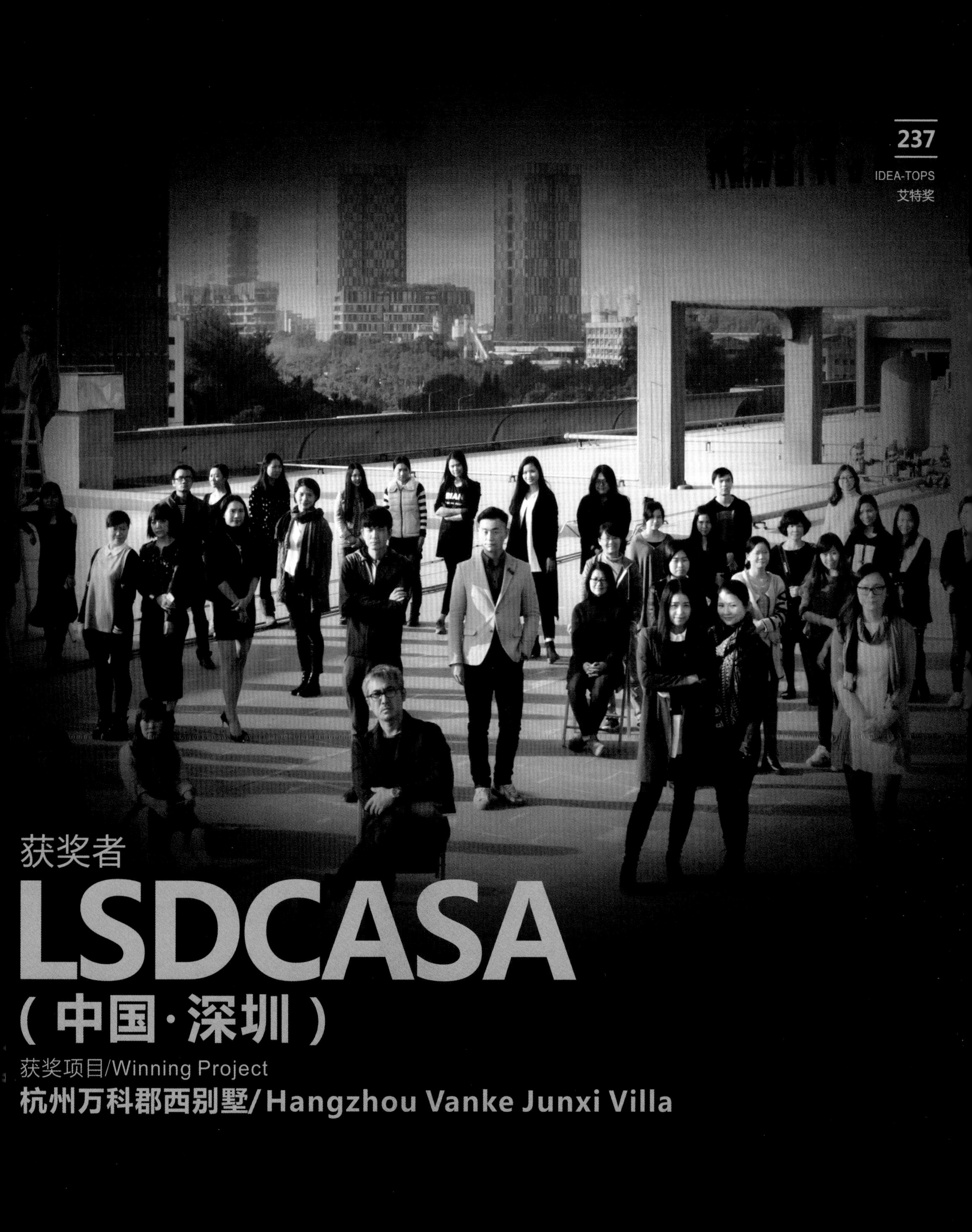

237
IDEA-TOPS
艾特奖

获奖者
LSDCASA
（中国·深圳）

获奖项目/Winning Project
杭州万科郡西别墅/ Hangzhou Vanke Junxi Villa

获奖项目/Winning Project

杭州万科郡西别墅
Hangzhou Vanke Junxi Villa

设计说明/ Design Illustration

万科杭州良渚文化村食街里"食堂"的"打饭"方式和"搪瓷缸",以及安藤忠雄设计的美术馆,让我们看到万科和建筑师们的企图。关于建筑的记忆,不只是来自一栋栋建筑,更是来自一个"地方"。万科在"营造地方",唤醒庭院和邻里的记忆,唤醒文化和之于生活的价值,奋力构建美的可能,而这些正是现今中国建筑师最根本的使命和责任。良渚文化村的建筑者知道,建筑师是文明的守护者。我们的室内设计正是在这一前提下开始,我们心生喜悦。

营造"地方"

杭州地区历史中归纳为江南,南宋是华夏文明最后的高潮,拥有比唐代更有文化教养的阶级。古人美善不分,孔子讲,"君子惠而不费,劳而不怨,欲而不贪,泰而不骄,威而不猛。"而表现形式和美学趣味一直在演变,到北宋的"无我之境"再到南宋的"有我之境"可类比西方印象派的源起,从描摹世界转向关注世界之于主观的感受和发现,表达出某种较为确定的诗意、情调,诗素以含蓄为特征,有诗意的画,所谓"含不尽之意见于言外。"

设计传承温润如玉,清风傲骨的知识分子情怀,舍宋词中西湖的妍姿幽态,取中正平衡,结合创新,结果出人意表,却又和谐自然,不怨、不贪、不嗔、不骄。

行进空间

建筑不只是空间设计,不是量体,不是如何组织体积,这些是附属在重点之下,而重点是如何组织起整个行进过程,建筑只存在于时间中。——这是赖特强调的"走过建筑"的概念。

高且密的竹林小道,造成的静逸情绪延续到紧致花园,直到步入室内玄关,落地玻璃让室外花园成为借景,设计师在此,用深色木作,工整的线条,严谨静态的摆放,造成时间的停顿和空间的逼仄,为进入客厅和餐厅的舒畅宽阔作铺垫。

客厅墙纸设计成有外墙质感,进一步把花园泳池和客厅空间,在感受上变成一体。行进至家庭房区域,用"亭台"的设计语言模糊了空间的界线,让室内室外再一次相互交融。

下行到负一楼,围合和包裹,仅留远景墙面的间接自然光,虚实、对比的书画手法,在空间内体现,制造空间的安宁和温暖的同时,在审美层面连接传统意境。

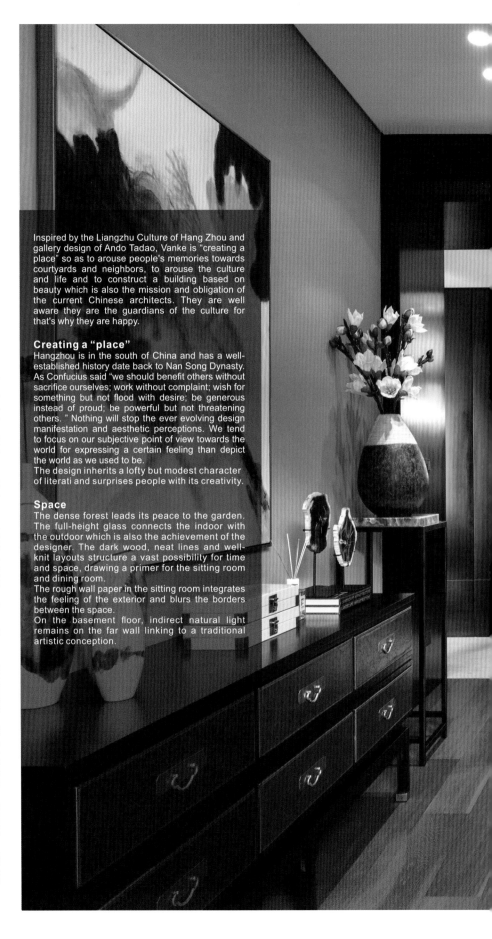

Inspired by the Liangzhu Culture of Hang Zhou and gallery design of Ando Tadao, Vanke is "creating a place" so as to arouse people's memories towards courtyards and neighbors, to arouse the culture and life and to construct a building based on beauty which is also the mission and obligation of the current Chinese architects. They are well aware they are the guardians of the culture for that's why they are happy.

Creating a "place"

Hangzhou is in the south of China and has a well-established history date back to Nan Song Dynasty. As Confucius said "we should benefit others without sacrifice ourselves; work without complaint; wish for something but not flood with desire; be generous instead of proud; be powerful but not threatening others. " Nothing will stop the ever evolving design manifestation and aesthetic perceptions. We tend to focus on our subjective point of view towards the world for expressing a certain feeling than depict the world as we used to be.

The design inherits a lofty but modest character of literati and surprises people with its creativity.

Space

The dense forest leads its peace to the garden. The full-height glass connects the indoor with the outdoor which is also the achievement of the designer. The dark wood, neat lines and well-knit layouts structure a vast possibility for time and space, drawing a primer for the sitting room and dining room.

The rough wall paper in the sitting room integrates the feeling of the exterior and blurs the borders between the space.

On the basement floor, indirect natural light remains on the far wall linking to a traditional artistic conception.

NOMINEE FOR BEST DESIGN AWARD OF VILLA

最佳别墅豪宅设计提名奖

Satoshi Kurosaki（日本）

获奖项目/Winning Project

Le49

设计说明/ Design Illustration

位于神奈川镰仓市的镰仓山上，它俯视着相模湾的美景。该项目的委托人是一对夫妻，他们曾住在东京闹市区的高层公寓里，来到这里之后就对绿树环荫的景色一见钟情，于是决定搬进来。对建筑痴迷的丈夫在英国工作时曾览尽欧洲的建筑，打算在此建一幢住宅并兼纳妻子的一间工作室。因此，我们打算创作一个现代并集日本美学为一身的独特建筑，也在此欢迎远方来的外国朋友。

房子是外表涂了一层白色光化漆的长方形建筑。这些错列的长方体所打造的重叠效果不仅具有现代感，还包涵了某种东方神韵。当你面朝大海，从狭窄的楼梯往下走的时候，屋子入口处的底层架空柱将映入眼帘。这也是别墅的一大特色。

一楼工作室前的院子中所种的树，在一定程度上保护了屋内的隐私。处于卧室一角的浴室，被落地玻璃隔开，带来一种非同寻常的体验：天然的绿荫美景环绕，宛如身处度假酒店般令人放松。与一楼的私密性相反，二楼是一个宽敞开放的空间。此处的玻璃尺寸是经过精心计算的，以便欣赏到周围的绿树和海景。储藏室、水槽和其他设施都被巧妙地嵌入墙壁之中。

这幢别墅最引人入胜的地方在于其钢铁和木头构成的五边形横梁，是非常大胆的设计。尽管如此，其中也不乏优雅的细节之美。滑门装置使这间奢华别墅内的住户随时与户外的大自然互动，同时也令日式美学得到升华。

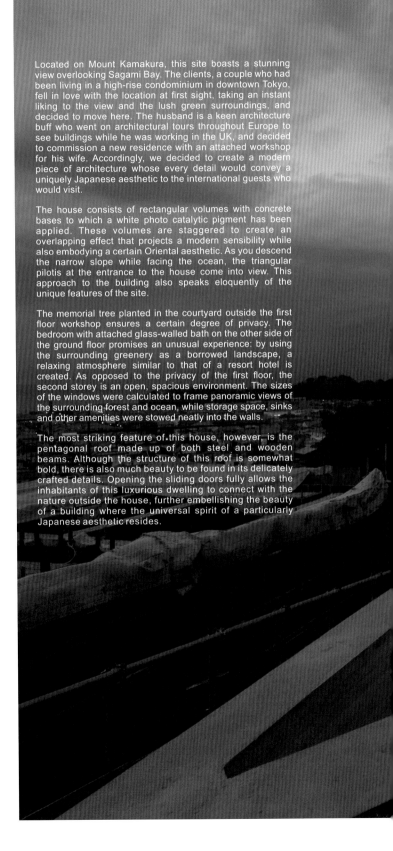

Located on Mount Kamakura, this site boasts a stunning view overlooking Sagami Bay. The clients, a couple who had been living in a high-rise condominium in downtown Tokyo, fell in love with the location at first sight, taking an instant liking to the view and the lush green surroundings, and decided to move here. The husband is a keen architecture buff who went on architectural tours throughout Europe to see buildings while he was working in the UK, and decided to commission a new residence with an attached workshop for his wife. Accordingly, we decided to create a modern piece of architecture whose every detail would convey a uniquely Japanese aesthetic to the international guests who would visit.

The house consists of rectangular volumes with concrete bases to which a white photo catalytic pigment has been applied. These volumes are staggered to create an overlapping effect that projects a modern sensibility while also embodying a certain Oriental aesthetic. As you descend the narrow slope while facing the ocean, the triangular pilotis at the entrance to the house come into view. This approach to the building also speaks eloquently of the unique features of the site.

The memorial tree planted in the courtyard outside the first floor workshop ensures a certain degree of privacy. The bedroom with attached glass-walled bath on the other side of the ground floor promises an unusual experience: by using the surrounding greenery as a borrowed landscape, a relaxing atmosphere similar to that of a resort hotel is created. As opposed to the privacy of the first floor, the second storey is an open, spacious environment. The sizes of the windows were calculated to frame panoramic views of the surrounding forest and ocean, while storage space, sinks and other amenities were stowed neatly into the walls.

The most striking feature of this house, however, is the pentagonal roof made up of both steel and wooden beams. Although the structure of this roof is somewhat bold, there is also much beauty to be found in its delicately crafted details. Opening the sliding doors fully allows the inhabitants of this luxurious dwelling to connect with the nature outside the house, further embellishing the beauty of a building where the universal spirit of a particularly Japanese aesthetic resides.

245
IDEA-TOPS
艾特奖

NOMINEE FOR BEST DESIGN AWARD OF VILLA

最佳别墅豪宅设计提名奖

孟也（中国·北京）

获奖项目/Winning Project

路
Road

设计说明/ Design Illustration

之所以将此项目命名为"路"，缘由来自于设计师对客户的真切了解及美好的祝福，两位主人相濡以沫、相互依偎与跟随、无论平坦曲折，一路走来，从年少到白发，共同建造了属于这个家庭的和谐与美好，是空间中需要表达的核心价值。这是两位受过高等教育、有过经商经历的学者组成的家庭，他们在退休前需要一套属于赋闲后养心修性的居所，在这里他们可以完成自己欣赏的文学、艺术、品茶论道等心愿，并也盼望着在美国留学的儿子学成归国，带着未来的儿媳和孙子或孙女共同陪伴着度过自己最重要的日子。

空间中重新规划的动线中，在满足了高效的使用同时，恰到好处地体现了这条路的幽远、曲折、起伏与回转，移步易景，体现了进门后内花园的概念感受。设计师在动线设定中有意拉长人的进深运动长度，欣赏沿途风景，曲转之间游走于计划好的视觉感受中。而点睛之处是架在挑空厅之间的桥，在营造了空间的情趣同时，也形成了一条通向住宅内部的捷径，是纪念男女主人以智慧打造人生捷径，走向成功的经历，紧扣项目主题。桥的尽头是通向客厅的上下楼梯等回转空间，也是建筑形态最为曲折多变的线条枢纽处，在不同的角度，会有着不同的结构线条及光线变化影响着没有方向感的你。

在一层的空间规划中，曲径通幽的东方园林理念与路、景观点结合，突出结构力量感，弱化模糊方向感，营造了一个情趣路径及视觉感受空间，让人在路过时有不同寻常的感受。挑空中，中国设计师原创的铁质装置被当作吊灯，呼应空间结构线条，路过桥时的近距离感受更加立体。

下沉庭院被改造为餐厅，引入阳光、绿植，主人的茶艺、甜点、正餐、音乐欣赏、静心阅读、家人谈心均被安置此处，甚为温馨，成为完成所有家庭情趣休闲文化同时兼具主餐厅的功能空间。卧室中，高挑的空间给了云朵灯更多飘摇的愿望，日本设计师给灯赋予了东方特有的细腻感受。空间主要家具全部为中国艺术家们精彩的作品，充满东方审美情趣，并不时结合西方印象，与国外设计师的小件配饰家具呼应，成为空间国际化印象的重要组成部分。

整个设计中，设计师孟也以现代空间打造的手法，融合中西方感人的审美情趣，赋予空间模棱两可的多元素风格感受，块、面、体、形一气合成，使用上更让空间充满情趣、和美，达成了中国人居最美好的愿景！

The reason why named this project as Road lies a wonderful wish of the designer for the client. The couple of this house cherish and love each other throughout their lifetime no matter what difficulties are, who built the family on joints efforts.

They are well educated and retired from business world, wanting a peaceful dwelling for cultivating their minds and knowledge, expecting to spend their most important time with the offspring.

The rearrangement of the garden not only realizes its effective function, but also manifests the twists and turns of the road. Therefore, the guests will have more time appreciating the outlook along the way. What amazes us is the suspended bridge across the void space leading a shortcut to the sitting room which embodies the successful paths the owners had took. Orientation the house such as lines and light makes it easier to find your way.

On the ground floor, the structural roads in the eastern garden touch the concept of project and incite unusual feelings of us at the same time. A design metal chandelier works in concert with the spatial sense.

The sunken garden is used as leisure space while introducing sunshine, plants, cuisine, music and books both for family chitchat and dining room.

Cloud shaped lamp from a Japanese designer invests the bedroom more eastern charm. Most of the furniture is crafted by Chinese artists, but from time to time you will discover some western home accessories composing a dynamic cosmopolitan ambience.

It consists of eastern and western appeal in a modern touch where is a dream house for the Chinese.

NOMINEE FOR BEST DESIGN AWARD OF VILLA
最佳别墅豪宅设计提名奖

深圳市矩阵室内装饰设计有限公司

获奖项目/Winning Project

重庆万科城LP6别墅
Chongqing Vanke LP6 Villa

设计说明/ Design Illustration

低调奢华的空间，用现代简洁的手法表现。
整体的功能分区动静分明，流线畅然。负一层娱乐派对，整层的互动与开放；到一层的休闲起居和楼上的学习、休憩，落落大方的格局组合，让生活更是从容。演绎出精致而丰富的现代精英生活，更是值得向往。
干净而凝练的表达，有力却不失细节和质感。同色系、同纹理的组合，让冰冷的空间，注入温暖的语言，散发现代都市新贵尊贵的气质。在材质和色彩的搭配上，十足的考究。尼斯木立体的纹理搭配黑镜钢的力量，再配以鳄鱼纹理的绒布硬包，低调干练，传达精致的感情。

This space implies extravagancy from its interior

performed in a concise and modern sense.
It is divided into static and dynamic functional area. For one thing, the open and active basement level is for partying; for another, the ground and first floor area is for living and studying, taking on an exquisite and diversified elite lifestyle.

It is refined in style but flourishing in details. The same color scheme meets the material of same texture transforming the lifeless space into a warm home, full of modern and distinguished touch, such as the combination of wood texture, black mirror plus crocodile patterned cushions.

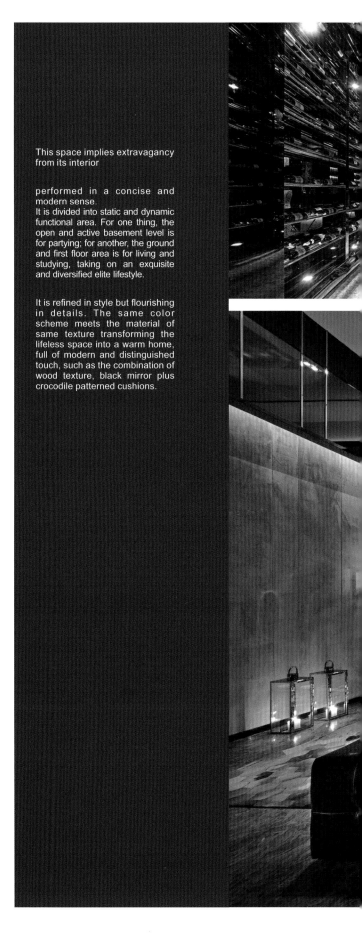

NOMINEE FOR BEST DESIGN AWARD OF VILLA
最佳别墅豪宅设计提名奖

Stefan Meyer（英国）

获奖项目/Winning Project

巴厘岛度假别墅
VILLA IN SEMINYAK BALI

设计说明/Design Illustration

开放式厨房和餐厅的高天花板上，吊着自然元素的装饰物，柔和自然的感觉为整体环境增色不少。客厅及餐厅结为一体,开放式客厅让蓝天白云充当天然的背景,大型舒适的沙发,以简洁的表现形式来满足人们对空间环境那种感性的、本能的和理性的需求，这是当今国际社会流行的设计风格——简洁明快的简约主义。

设计师依循共性与个性的线索，结合"实用艺术"的原则，于是便有了以下与大家分享的设计案例——巴厘岛的低调奢华。西方女人的水彩照片、木质的门框架……令巴厘岛酒店拥有迷人的风韵。休息区，优雅的白色空间配以长形L形沙发，让身心彻底放松。

餐台采用做旧的长型餐桌，黑色条纹瞬间让空间产生几何感，镜子的摆放映出了对面的景色，同时也为广阔的空间做出一幅流动的画。

这里的家具由实木制作而成，在厚重质感的同时，利用色彩的搭配，起到了良好的视觉冲击作用，经过多层叠加后也会带来意想不到的奇妙效果。巧妙的布局、精致的光线、淳朴的材料、精准的线条和利落的细节，为一个家铺陈出诗意的情调。白色迷人的北欧风格装修设计，任何人看到这座复式楼都会感到惊叹。无可挑剔的白色勾勒出一道道柔和的线条，创造出清爽的气氛。明亮的灯光使得屋内每一处都可以熠熠生辉，而宽敞的露台以及迷人的风景也是令人叫绝。

别墅的一层围绕泳池进行设计，无垠的游泳池里倒映着太阳和天空，天花板风扇调节着稳定而自然的气流,此外还配置有可供6人坐的大型餐桌和非常舒适的沙发。

除了维多利亚式镀金大镜子以外，Stefan并不过多追求繁复设计，竭力使设计最简。大部分具有时代印记的复古经典家具在白色墙壁的反衬下显得历史感十足。Stefan习惯以色彩划分层次，这次，他将传统与古典格调混搭明朗色彩，希望这个明媚的别墅中充满各种各样的色彩，特别是那些心仪的色彩要毫无保留地运用在家中。自然光固然好，却要运用得有张有弛。你可以将它当作一种光的工具，利用每天不同时间的日照角度和幅度，为特定的角落、细节、材料和纹理打上一些阳光，突出它们的美。

It is designed for a private business man as a holiday home retreat that could display his special collection of art.

The brief was to create a space that would capture the sea breeze as it flows through the property.

This two story villa has 6 bedrooms, 7 bathrooms, a games room, spa and mini gym.
It was designed to accommodate and entertain 2 families holidaying together. As such a huge open-air living space and outside dining area were created along with a large roof top terrace and two separate gardens. The idea being that there will always be a space for different activities to take place comfortably at the same time or just a place to disappear and read a book quietly.

The style chosen is that of fresh unpretentious quality - Contemporary 'Bali Funk'.

260
Best Design Award of Art Display
最佳陈设艺术设计奖

艾特奖
最佳陈设艺术设计大奖
BEST DESIGN AWARD OF ART DISPLAY

IDEA-TOPS
INTERNATIONAL SPACE DESIGN AWARD

获奖者
陈俊良
(中国·台湾)

获奖项目/Winning Project
原来台湾/ Original Taiwan

获奖项目/Winning Project

原来台湾
Original Taiwan

设计说明/ Design Illustration

原来台湾是第一个将"台湾原住民生活美学"作为主题所策划的展览，跳脱以往强调差异与异国情调的美学认知，具有原住民生活美学产品的示范性与指标性意义。原住民文化与美感不再以退入差异符号的方式做为晋升被主流世界看见的方式，而是可以进入你我日常生活的美好生活态度。其中更以利用台湾原住民的传统有机素材，如：漂流木、石板、陶土、树皮布、竹、月桃、香蕉丝等，使其不只是温暖、亲近有触感的有机材，更承载了源流不断的文化生命与故事：愉悦劳动、传承使命、追求幸福。展览以生活居室呈现，使展品更有生活化的想象，将内建的传统生活与美感，外显于现代生活产品。例如客厅区的石板边桌、树皮吊灯、原住民色彩的配色壁饰；餐厅区的漂流木桌椅与餐盘；卧室区的冬暖夏凉月桃席、织品抱枕等，让深度艺术融入简单生活之中。此外原来台湾展览，与多位原住民艺术家合作，如：拉黑子、达立夫(漂流木创作)、尤玛、达陆(织布)、巫玛斯、金路儿(琉璃珠)等，结合原住民工艺与现代设计，将深度艺术融入简单生活之中。使用各种有机自然素材，创作贴近日常兼具原住民之美的生活用品，成为台湾原住民的新生活美学的最佳范例。

Original Taiwan is the first exhibition revolved around Living Art of Taiwanese Aboriginals. It is different from other exotism introduction instead of illustrating the life of them, a life that we are familiar with. There are organic materials of Taiwanese Aboriginals, for instance driftwood, slates, pottery clay, tapa cloth, bamboo, alpinia zerumbet and banana tree bark which carry cultural stories with endless vitality. The exhibition is rendered in a life style.

Moreover, Original Taiwan collaborates with several aboriginal artists, for example Rahic Talif, Yuma Taru, Umass Kinlourie. They take advantage of the natural materials blending conventional craftsmanship with modern design. Design for life is all their aim.

IDEA-TOPS
艾特奖

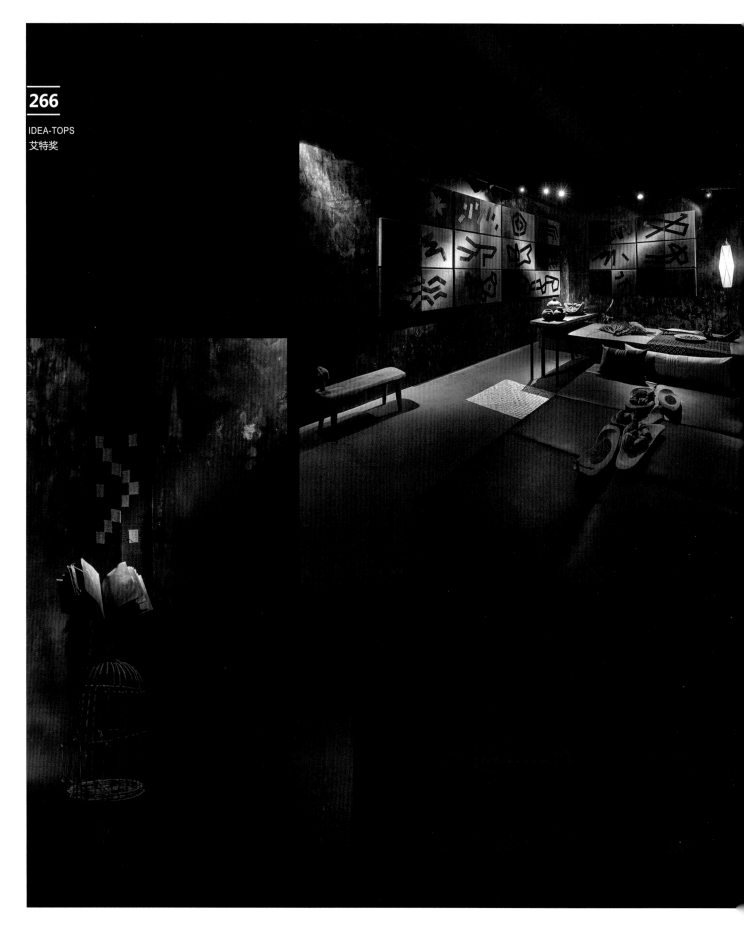

NOMINEE FOR BEST DESIGN AWARD OF ART DISPLAY
最佳陈设艺术设计提名奖

深圳市天琢轩设计有限公司

获奖项目/Winning Project

江南1900
East China 1900

设计说明/ Design Illustration

本案设计回归自然，遵循着自然的规律。集合了充满生活蕴意和时间年轮的普通生活器皿，将它们重新组合。利用不同的表现手法创造出不同的意味，展示出中国传统文化。借鉴的是中国文化中所说的"道法自然"，讲究的"因借"的观念，不做作，不执著于外在的形势，强调的是"得意忘象"，不执着于"象"而要抓住"意"。

Return to the nature and comply its rules is the concept of this project. To collect all the daily wares that rich in history and stories, it implies a multiple of meanings by different arrangement. We draw lessons from ancient wisdom "follow the nature" that we should not emphasize on the appearance but keep hold of the substance instead.

NOMINEE FOR BEST DESIGN AWARD OF ART DISPLAY
最佳陈设艺术设计提名奖

山景空间创意有限公司（台湾）

获奖项目/Winning Project

东方威尼斯 苏州中式样板房
Eastern Venice Suzhou Show Flat

设计说明/ Design Illustration

苏州，一座水色盈溢的古老城市，具有绝佳的水乡风景——如意大利威尼斯的地理特性；同时，悠久历史与现代发展，使此地于中西交流中，更呈现古今对话的可能。故此，特以湖水色泽为底，继之传统线条与现代材质的变化，铺就本案的美学元素。

踏入玄关，取材自知名建筑师莱特（Frank Lloyd Wright, 1867-1959）的繁复窗花映入眼帘，装饰主义的流利线条，与对口鞋柜的金箔花样遥相呼应，体现东西元素的戏剧张力；几扇鎏金窗花深嵌壁面，为客厅点缀古韵之余，亦成为串连视觉的利器，设计者进一步以镜面不锈钢天花的反射效果，转化了空间比例，增强了大器氛围；延伸线条起伏，餐厅以出风口串起内凹天花板，明晰着客厅餐厅界线的同时，亦使视觉开阔、倍显豪宅气势。

于色彩方面，以湖绿为底，将传统元素与现代工艺紧密结合，客厅窗帘选用明黄跳色，卧室选以不同层次的草绿与黛绿等，于古意盎然的廊室内，体现"中西混搭"——如马可波罗远渡重洋抵达中国，与忽必烈大汗把酒言欢、相互馈赠的和谐景致。

Suzhou is a historical city known for her extraordinary watery landscape just like Venice, and also advance in modern technology. This project merges the color of the lake into its background and combines traditional look with the modern materials.

As you entering the house, the lavish window decoration at the entrance-hall comes into view well joint with the art deco style and gold foil on shoe cabinet, reflecting both eastern and western elements. Gliding windows built-in the wall not only embellish the space with classical touch but also scale up the spatial dimension through the mirror. The concaved ceiling in the dining room establishes a boundary from the sitting room significantly underscoring this grand house.

As for the color scheme, base color is as green as the lake, binding classical factors and modern craftsmanship together. The curtain in sitting room is yellow, color of bedroom goes with levels of grass green and dark green, a mix and match style of the east and west.

NOMINEE FOR BEST DESIGN AWARD OF ART DISPLAY
最佳陈设艺术设计提名奖

杨柏林（中国·台湾）

获奖项目/Winning Project

山石山象工作室
Stone Studio

设计说明/ Design Illustration

自然与人文、现代艺术与古董文物、东方风格与西方元素所和谐建构的世外王国。这幢沿着山坡所建的建筑物保留了许多原生的参天大树，"树木仿佛可以触摸我的身体，摸到我的微笑，摸到我的悲伤。"杨柏林居家与工作室像是自然空间的剧场，满溢自然的气氛。"若我是鸟，这个建筑就是我的鸟巢，让我有能量在喜欢的空间里生活，进行创作。"十几年过去，它有机地延伸，依然持续与自然对话着。

"我喜欢古老的东西，古董是我的百科全书，许多创造性语汇，深置其中，有种神秘的张力，有生活感动又有创造性的神秘喜悦。"

"艺术品或古董她们有相同的语言，若品味在同一个纬度，她们自己会聊天。会有生生世世的缘分。"空间里有棵树，像是在室内，但是却是户外。杨柏林尊重自然，他相信自然会回馈给他全新的、启发性的、创造性的灵感。

杨柏林认为定居在山林里是这个地点对他的召唤。将自己多年对生活的哲学与美学的理解投入于这个自己规划设计的建筑之中，"树木的种类是我定的，空间里的艺术品是我找的……这个家所反映的是我沉淀过的思维。理解我的生命深度，不能只看我的雕刻，必须加上我的画、我的文字、我的生活、我的家、我的收藏……全部一起，才能呈现我完整生命的特质与境界。"

The idea is to build Xanadu encompassing nature, modern and antique art along with the east and west culture. It is built along the mountain slope surrounded by numerous primitive towering trees. "It seems the tree could touch me, my smile and sorrow." He said.

Here is like a space of the nature. "If I'm a bird, then this is my nest for living and working. I like antique stuff. They are my encyclopaedia that touches and inspires my life."

"Art and antique have their own language that they can chat if they have something in common" quoted. The designer respects and believes in the nature which returns him creativity as a generous gift.

He thinks this mountain recalls him to dwell in where he could realize the philosophy for life and design developed after these years. "I choose the tree and art pieces for this house. To understand me, you have to see my sculpture, my painting, my writing, my life, my house and my collection. All of these will reveal me as an integrity."

文特奖
最佳公寓设计大奖
BEST DESIGN AWARD OF APARTMENT

IDEA TOPS

INTERNATIONAL SPACE DESIGN AWARD

获奖者
琚宾（中国·深圳）

获奖项目/Winning Project
蜗居27m² / Snail House of 27m²

获奖项目/Winning Project

蜗居27m²
Snail House of 27m²

设计说明/ Design Illustration

这是一个公益项目。通过27平方米的loft空间，规划出集居住、工作双重功能于一体的自由职业者的向往之地。这种向往与实施，能让更多的自由职业者，通过当下科技与互联网的平台工作和学习，避免交通的拥挤以节约所耗费的珍贵时间。

希望这能唤起社会的不同角度认识，以设计师的方式表达一种观点和情怀。材料是空间表情的最终诠释物。白色本身的中性特质，让空间舒畅地呼吸并拥有更多的自由，也为留住时间的年轮、情怀的印记打底。其颜色本身便能让小空间有着更大的视觉想象，同时也承载了简与素所能衍射出的建筑的本质。剥离掉装饰，让空间围合出独特的空气，并与社会保持着适当的距离。

用至简近道的方式来表达混合功能的多样性，在明晰可辨的逻辑下，寻找丰富的、快乐的空间本质魅力。

光，作为这个设计本体的主要材料，在空间中以多角度多方式的呈现手法。留住光的同时，也是留住了时间，留住自我观照时的那份宁静。

This is a social project, a snail-house of 27m² is a loft design to integrate freelancer's living and working. The purpose of this project is to rethink an alternative life of freelancers, who lives and works in the context of an internet age.

The designer hopes to evoke a different perspective of the society through a designer's interpretation. Material is the ultimate means of spatial expression. White is a neutral color, letting space to breath and enjoying freedom. White inherits the nature of architecture that is simple and modest.

This project displays the multiplicity of programs with simple means. It is through this clear logic that the fundamental nature of space can be enjoyed and discovered.

Light is another material that used in the design in multiple ways. To capture light is to capture time and the self in work and life.

286

IDEA-TOPS
艾特奖

NOMINEE FOR BEST DESIGN AWARD OF APARTMENT
最佳公寓设计提名奖

Teresa Sapey（西班牙）

获奖项目/Winning Project

波尔多LOFT
LOFT BOURDEAUX

设计说明/ Design Illustration

它是法国波尔多工业区的一个车库改建成的，公寓焕然一新地被分割成两个主要区域：一个是招待访客的公共区域，另一个是居住空间。前者包括一个厨房、酒窖和客房，主人的私人空间则包括一个大躺椅，与书架和庭院相连。室内泳池、健身房和盥洗室延成一线，与卧室和更衣室比邻而居。楼上设有台球桌，俯视着整个公共区域并与躺椅所在的休息区遥遥相望。
庭院里摆放了Teresa Sapey工作室出品的红色、橘黄色和酒红色的壁画以及一张木桌。花架和黑色铁艺座椅是出自西班牙品牌Los Peotes。地板是柚木制成。
休息区中的躺椅是特别定制的。客厅内的书架和地毯是由Teresa Sapey设计，电视背景墙上的油画是出自著名的艺术家Alberto Acinas之手。此外，还可以见到Roche Bobois品牌的台灯、花瓶和靠垫以及Vitra Eames品牌的胡桃木凳子。
小阁楼上的宽敞休息区内设有台球室，在这之下的空间是用餐的地方。公寓保留了原有建筑金属的横梁结构，黑色的油漆更突出了它的存在感，又与白色墙壁形成视觉反差。
通往台球室的台阶是橡木所制，与地板相辅相成。阁楼上的壁柜既可以摆放书籍也能储物，白色的设计增加了空间感。这些都是由Teresa Sapey设计。
厨房的设备上用法语写着巨大的"我讨厌做饭"的字眼，这是公寓主人开的一个玩笑。墙上大大小小的圆洞设计是用来摆放酒品的。

Situated in a former garage in a Bourdeaux industrial district (France), this house was fully refurbished and divided into two main areas: a public one for interaction with visitors and private living space. The first comprises a kitchen, cellar and guest room. The owners' private space contains a large lounge linked with book-cases and an inner courtyard. The indoor swimming pool links the gym to the bathroom, which in turn leads to the bedroom through a dressing room. A space with a snooker table on the upper floor overlooks the lounge.

In the courtyard, we find murals in red, saffron and wine tones made by Teresa Sapey's Studio, as the table with wooden top. Planters and black metal chairs are from a Spanish Brand named Los Peñotes. The floor is made of teak.

In the lounge, we find a tailor-made sofa with chaise lounge. The storage unit and the carpet have been designed by Teresa Sapey. Over it, we find an oil painting from the famous artist Alberto Acinas; table lamp, vases and cushions from Roche Bobois; turned walnut stools, designed by Eames for Vitra.

The spacious loft lounge, with billiards room located in a high space or attic, under which stood the dining. The old structure of metal roofing and cladding has been painted in black to highlight their presence and provide a visual counterpoint to the white walls.

Detail of the staircase leading to the billiard room with floating steps in oak matching the pavement. The movable wall that runs along the wall of the room serves as a library and solves the storage. Its white finish helps to lighten its volume. Everything is designed by Teresa Sapey.

On the kitchen furniture was written in French with vinyl lettering 'I don't like cooking', a joke to the Loft's owner. Round coated metal cavities of different sizes were made on the walls to place the bottles.

NOMINEE FOR BEST DESIGN AWARD OF APARTMENT
最佳公寓设计提名奖

I29 Interior Architects（荷兰）

获奖项目/Winning Project

家11
home 11

设计说明/ Design Illustration

这里原是阿姆斯特丹德派普区的一个车库，现被改造成一个自然光线极佳的宽阔公寓。一楼230平方米的偌大空间运用了简单的材料与色调，反复出现粗犷的长条橡木板，与白色墙壁、黑色台面和灰色调地板形成鲜明的对比。

在量身定制的厨房中，一个巨大的木制拉门可以覆盖整个储物柜，与前方的黑色灶台形成强烈的色彩反差。嵌入式的壁橱和壁炉也具有同样的特色。卧室和洗手间的入口是内嵌的落地高木门，与墙壁融为一体也更加隐秘。

在客厅和主卧室之间，恰巧是户外天井的位置，这样有利于加强室内和室外的互动体验。此外，公寓内20平方米的青苔式样的地毯是I29公司亲自设计的手工编织品。充裕的自然光照射在柔软的绿色和米黄色相间的表层时，让室内之人仿佛置身户外。

A formerly garage space in Amsterdam's area De Pijp, turned into a spacious house naturally lit by large roof lights. The interior with a generous 230m² on one floor level is finished in a simple material palette. The repetition of rectangular rough oak wooden surfaces is in great contrast with the stark white walls, black surfaces and grey cast flooring.

The custom designed kitchen includes a large wooden sliding door to cover integrated storage areas, with a contrasting black cooking island in front. Built-in cabinets and a fireplace have the same characterics and contrast in materials. Wooden walls from top to bottom with built-in doors are marking the entrance to the more private areas such as bed and bathrooms.

Outdoors is a patio in between the living and master bedroom. In order to heighten the outdoor experience and to connect inside and out, I29 interior architects designed a 20m² hand knotted carpet with a natural mossy pattern. The excess of natural light in combination with the soft layer of green and beige resembles the outside experience while being inside.

NOMINEE FOR BEST DESIGN AWARD OF APARTMENT
最佳公寓设计提名奖

廖奕权（中国·香港）

获奖项目/Winning Project

树影下之悠然
Leafy Shade

设计说明/ Design Illustration

树影下悠然徜徉
想象安躺于树下，树影摇曳，间或于枝叶隙缝间，隐隐透进自然光线，温暖人心；间或吹来阵阵凉风，沁凉透入心扉。万物不过虚无，无穷想象力却不受限于任何羁绊，驰骋于天地之间。

树影婆娑，枝叶随风摇曳
玄关处，不见任何植物或盆景，却处处觅得自然气息。木制屏风、叶子形状修饰，拼凑起来仿似错落枝叶。放眼远望，屏风自入门处连绵不断，弧形线条至玻璃窗戛然而止。明亮无压的空间，辅以天花之镜面装饰、质感朴实原始的茶几，令人仿若置身树荫之下悠然徜徉，感受美不胜收的自然光景。放空身、心、灵，一扫生活的烦扰。

风光明媚，满目尽是自然
入门即见的厅堂以暖色调温暖人心，木色与澄黄调铺饰墙身。以招待为主的偌大空间毫无顾忌，放任添注金、黄的鲜明色调；又借现代语言将单位塑造如同精品酒店，一室气度浑然天成，却不落俗套，亮眼色调的点缀适可而止，亮不张扬。客厅一隅，茶几、沙发、地毯、灯具、屏风，据以营造之弧形线条大度、圆润成为当眼主角；饭厅反是沉稳平实线条当道，让云石、镜、木等异材质共同拼凑，互为表里。一切起源于想象，居于内，又形同处身自然：纵使形态、个性不一，亦该于同一环境共融，和平共存。

目不暇接，误闯森林深处
一转角、一拐弯，富艺术感的装饰品无处不在，不着痕迹地提升空间，又恰到好处地留白。拐过让人心旷神怡、悦目地下厅堂，不期然令来者放轻脚步。梯间，玫瑰金不锈钢屏风象征过渡。设计师把握操控颜色的魔法，深浅有致。浓淡得宜之色调运用，不徐不疾将厅堂焕发神采的用色逐步加深。上层的休息区域率先以浅色木地板营造温润自然感，房间木色充斥，辅以变化细腻的灯光效果，铺陈柔和卧眠气氛。

The principle concept established for this project at the Mandarin Oriental in Macau, is to embody tranquility at its essence through capturing the elements of nature. The Tree is symbolic of being protective, sheltering and being able to nurture life-giving energy. The idea is to allow its occupants to enter a nature-inspired living space, where they could feel relaxed and safe within the comfort of their own home.

Walking through the entrance door reveals a welcoming feeling of freshness with leaves are blowing and swaying towards the ceiling. Abstract tree branches encircle the living area are accentuated by the autumn leaves which project along the curved decorative feature. This curved feature along with the reflective ceiling feature recreates an experience akin to laying under a tree with reveals of light breaking though the tree canopy. Earthy tones and choice of natural materials enhances the natural feeling within the space.

The dining room reciprocates the natural style of the living room with added elegance. The long wooden dining table, capable of accommodating a large number of guests, directly joins into the polished marble structure with a concealed television which is available for viewing when hosting. Alongside is a fireplace installed on the wall, highlighting the contrasting material of the shelf. The stair features draws you in to continue the story as it unfolds towards the second floor. The rose gold finish brings life to the staircase. The handrail enhances the contemporary feel by implementing the elongated shape of tree stems. Adjacent to the living room, the ceiling of the second floor is designed with the pattern of abstract tree branches.

The wall hung grey marble TV wall extends along the corridor wall; the corridor opens up to a cozy corner for enjoying your favorite beverage. Contrasting with the material used on the wall, the warm wooden flooring covered with a stone shape rug captures the essence of nature. The master bedroom has been purposely designed in a deep brown color with a back wall feature designed to simulate tree trunks layered upon one another. As well as looking natural, the choice of colours and materials offers a sense of luxury in this area. Concealed behind the grey and white marble finished wall, the bedroom has been designed to the same extent as the master bedroom. The bed is framed by unique bed side cabinet features in a copper finish, whilst the lighting creates the illusion of light simmering through a tree canopy. The bedroom wall and wardrobe are covered with glass to visually enlarge the space and the geometric carpet adds texture and colour, giving a playful touch to the glamorous room.

297
IDEA-TOPS
艾特奖

NOMINEE FOR BEST DESIGN AWARD OF APARTMENT
最佳公寓设计提名奖

城市室内装修设计有限公司

获奖项目/Winning Project

线条山水
Line of Landscape

设计说明/ Design Illustration

本案坐落在拥有L形落地景观窗的顶楼住家，依山傍水的天然格局，成为本案设计的发想。

引用窗外的山河曲形线条，延展出室内的天花板造型。让户内与户外的氛围，能有线性的连接与融和,同时也削减了原有过多横梁的视觉障碍。

墙面实木拼板与反射明镜材料的使用，创造了动态的虚实空间感，也让原有单向的景观与采光，在室内有了新的观赏角度与乐趣。

以跨区域性的混搭选配,让原始有机与工业几何的强烈对比，带出了既随性又休闲、属于屋主的独特性格与样貌。

This apartment is not big yet claims an ample view. The clients' demand was a simple living space, two bedrooms and a living space big enough to accommodate every social gathering. The challenge of the design was to give the space an intimate comfort full of energy.

An L-shaped floor to ceiling window that accepts the natural view dominates the main living quarter. The space now appears as if there hangs a long scroll landscape painting where nature has placed a beautiful mountain and river in front of living space.

Entering from the lobby and moving towards the living space, one confronts a mirror reflecting landscape of the river. The corner mounted ingenious layout of a mirror provides an illusion. It is like a window penetrating into the woods as if it was totally surrounded by trees and an endless patch of greenery.

The ceiling has an undulating profile as it projects the reversed ground topography. The existing beams and columns seem to be confused with little order. We tried to re-configure the ceiling into a new form, which pays respect to the surrounding outdoor landscape while hiding the internal disoriented structural elements.

There is a sharp contrast between the ceiling pattern and partition wall. The wood ceiling pattern flows down and gradually disappears until it make a direct confrontation with the marble floor. The change is subtle and gentle. Before the wood material changes from ceiling to marble floor different wood materials were used and the texture changes from silky to fine and from fine to a striped pattern until the wood make a sharp change to marble.

The mixture of furniture style is confused yet stimulates an energetic dialogue of spatial interest.

304
Best Design Award of Commercial Space
最佳商业空间设计奖

文特奖
最佳商业空间设计大奖
BEST DESIGN AWARD OF COMMERCIAL SPACE

IDEA-TOPS
INTERNATIONAL SPACE DESIGN AWARD

获奖者
吕元祥建筑师事务所
(中国·广州)

获奖项目/Winning Project
广州天盈广场 / Guangzhou Top Plaza

获奖项目 / Winning Project

广州天盈广场
Guangzhou Top Plaza

设计说明 / Design Illustration

办公室位于广州市珠江新城黄金商业地段的大楼,旨在为国际商贸、信息科技及融资管理等国际企业提供一个开放进取的瞻远办公空间。设计突显项目之专业品位,有条不紊。

在狭小的空间,大气及细致并存,将建筑与艺术结合,展现刚柔并重,美观与实用兼备的办公空间。相比一般传统繁密的办公室规划,设计师创建了一个简洁又极具活力的空间,是国际视野的展示,对区内办公空间设计和营运文化的发展,奠定了重要的基础。

办公室的空间布局着重开放性与功能性交叉的和谐。设计师清晰地把办公室内的不同功能区细致划分,透过晶莹的玻璃,以流线直接的手法,凝造强而有力的实用性及空间感,尽显人性化及灵活性的考虑。

走进办公室接待处,便是平滑明亮的弧形接待台,仿如艺术品般迎接来宾;简约的条纹玻璃,展现出企业的国际品位;天花上深色玻璃的反射,令空间造型更见立体。会议室天花呈流线型,配合玻璃轴门,层次有序,贯穿彼此空间,构成利落的功能规划分布。在办公室中心位置的社交区,设有设计独特的酒吧桌,与接待台相呼应,为繁忙的商务增添趣味。

行政办公室的现代简约设计以窗前艺术品作点缀,突显管理人员的个人品位及精益求精的工作态度。工作间开放式的设计促进员工之间的互动,表现国际级的企业文化。

办公空间的整体布局细腻,以人为本。运用暖系色调融合橙色作为点缀,配合不同弧形线条,质感丰富,对比分明,体现了大都会的时代感。

The interior design embraces an international outlook through openness, quality and functionality. The sensibly-articulated space is a harmonious display of architecture and arts, rigidity and softness. The design concepts focus on the needs of service-oriented private equity funds business client. Our designers streamlined the office functions and created a crisp, clean space that is both dynamic and volumetric.

Visitors are greeted by an unconventional sculpted reception desk; behind this is a tunnel-shaped waiting area. The reception space is surrounded by custom-made glass panels with inlaid strips, creating a visual separation between the public zone and the office zones. In addition to these "statement" features, the entire space is subtly reflected by a tinted dark glass ceiling, creating a sensational welcoming effect for all visitors. These design concepts are extended into the conference rooms, where multi-layered ceilings generate a striking volumetric echo to the embossed glass panels. The open panel executive office projects excellence through modern simplicity.

The design intent is to draw potential buyers away from the traditional "headcount density" office environment and introduce them to the excitement of an international office culture.

The location of this office building at the energetic CBD of Guangdong Province is hugely significant. The location symbolizes the leading role of this development project, and defines and intensifies the building's international image across China.

The project's unconventional design utilizes a mix of vibrant colours and an open circulation flow to create a regional icon. The colour palette chosen for the backdrop of this dynamic office was a combination of orange and warm tones, giving depth to the intricately-layered three dimensional space and creating pleasant, delicate elements throughout.

NOMINEE FOR BEST DESIGN AWARD OF COMMERCIAL SPACE
最佳商业空间设计提名奖

Plajer & Franz Studio（德国）

获奖项目/Winning Project

几何题
Geometry

设计说明/ Design Illustration

"男人的数学"

"几何题"是男人专属的，它是一间位于柏林中心区的概念服装店。plajer & franz工作室把它打造成一间粗犷、时髦的并带着些许复古气息的男装店。"几何题"化身成一间格局考究的、怪异的数学教授的公寓，展示了他稀奇古怪的收藏，比如动物骨骼照片和稻草式的灯饰等。

"几何题"这个名字源于委托人对数学的热爱，plajer & franz以此为准则打造了这间概念店。它的面积不大，服装方面包括Gaspard Yurkievich、Irie Wash、Y3、Won Hundred 等设计师品牌，饰品有John Galliano以及Gijs Bakker、Arik Levy等当代本土设计师的家装饰品。

在柏林这样的城市，粗犷和时髦的店面未免乏善可陈，而plajer & franz工作室意欲为它营造一种全新的氛围。"几何题"想要以一个怪异的、拥有高雅品位的数学教授的公寓带给客户惊喜。

这位教授不但叫人捉摸不透，他对比例和氛围的把握还颇为讲究。他收集了各种与分析、探索和计算工作有关奇怪的物品：动物骨骼照片、稻草形的灯具。每件物品显示了对称与不对称，它的内饰也十分时尚，泥色的墙面，深色的木质地板和白色的橡木家具与男士的主题服饰店相得益彰。除了以上那些别具一格的搭配，它还融入了一些令人舒适的居家元素，如等候区的地毯和DNA灯饰。店铺中的陈设布局和商品展示的每一个细节都是经过精心安排的。

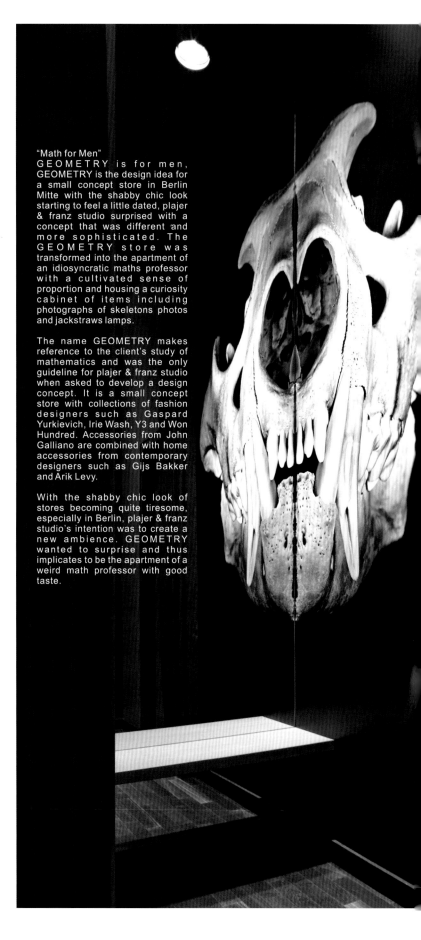

"Math for Men"

GEOMETRY is for men, GEOMETRY is the design idea for a small concept store in Berlin Mitte with the shabby chic look starting to feel a little dated, plajer & franz studio surprised with a concept that was different and more sophisticated. The GEOMETRY store was transformed into the apartment of an idiosyncratic maths professor with a cultivated sense of proportion and housing a curiosity cabinet of items including photographs of skeletons photos and jackstraws lamps.

The name GEOMETRY makes reference to the client's study of mathematics and was the only guideline for plajer & franz studio when asked to develop a design concept. It is a small concept store with collections of fashion designers such as Gaspard Yurkievich, Irie Wash, Y3 and Won Hundred. Accessories from John Galliano are combined with home accessories from contemporary designers such as Gijs Bakker and Arik Levy.

With the shabby chic look of stores becoming quite tiresome, especially in Berlin, plajer & franz studio's intention was to create a new ambience. GEOMETRY wanted to surprise and thus implicates to be the apartment of a weird math professor with good taste.

The professor is not only obscure but also has a cultivated sense for proportions and moods. He is collecting all kinds of strange items like skeleton photos, lamps that look like jackstraws having to do with analyzing, exploring and counting. Everything refers to the symmetry and asymmetry. Still everything is very stylish—the mud-colored wall, the dark wooden floor and the brushed white oak furniture and works well for a men's concept store. Despite of the unique atmosphere combined with homey elements like waiting areas with rugs and DNA lamps the design is not playing with the impossible. All aspects of the retail architecture and the associated presentation of the merchandise are planned and implemented right down to the last detail.

NOMINEE FOR BEST DESIGN AWARD OF COMMERCIAL SPACE
最佳商业空间设计提名奖

珠海天王空间设计有限公司

获奖项目/Winning Project

中泰家具博览中心
Zhongtai Furniture Expo Center

设计说明/ Design Illustration

本案属于一个旧楼改造项目，由于原建筑结构横梁和柱体的错综复杂无法更改或重建，设计师结合设计主题大胆地利用建筑主体的不利因素变成设计的元素，最后创造出最契合主题概念的空间。将不利因素变成有利，是本案设计亮点之一。

本案第二个亮点，是品牌文化与空间概念的契合，用现代的手法为中国传统文化赋予新的生命，此举实为弘扬中国的传统文化。本案中，中泰是以中国文化的元素作为企业文化的载体，博大精深的中国文化其实讲究的是一个"和"字，于是"和生万象"是中泰品牌文化的精髓。结合中泰品牌文化，设计师在众多中国文化元素中，筛选出六合（六和）、万字回纹、荷花、格栅等最契合概念设计元素，并大胆运用中国红作为主色，创造出震撼且富有感染力的视觉效果。设计上，重点在于运用现代形式及现代材料材质来表达中国传统"和"文化的内涵；通过VI企业形象与空间进行有机转换，做到品牌文化和空间设计的完美表达。

This design is an example of building reconstruction. Due to the complicated beams and pillars of the old building which cause great inconvenience to the restoration work, the designer turns this situation around and gives a new life to the space.

What is more, the designer extracts the brand culture from the proprietor and coheres with the spatial concept which enriches the marrow of our traditional culture. Zhongtai is a characteristic Chinese enterprise which values "coexistence and harmony". Consequently, numbers, characters and flowers that imply good fortune, such as "six", "fylfot" and lotus, are put into embellishment amenities. Red is applied in massive scale that is more striking and appealing for our eyes. As for the materials, they are implemented by a contemporary approach while interchanging brand culture in its core.

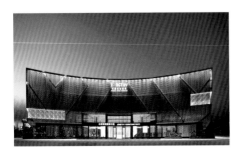

318
IDEA-TOPS
艾特奖

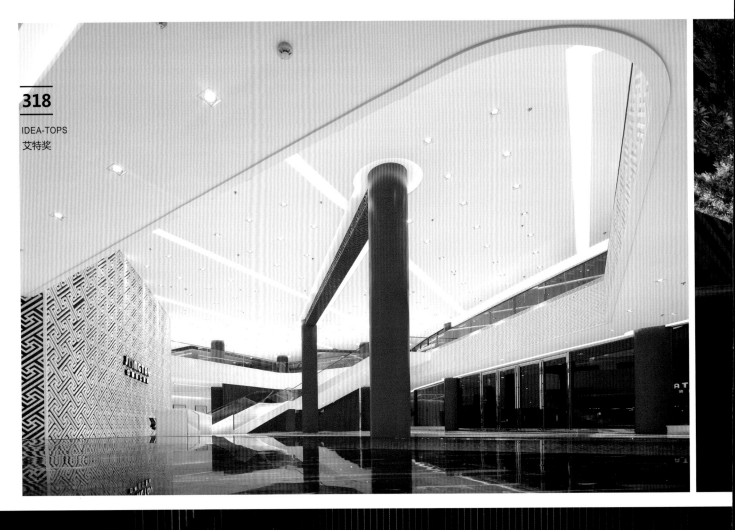

NOMINEE FOR BEST DESIGN AWARD OF COMMERCIAL SPACE
最佳商业空间设计提名奖

陈贻、张睦晨（中国·北京）

获奖项目/Winning Project
沈阳中铁丁香水岸售楼会所
Shenyang Zhongtie
Clove Bay Sales Center

设计说明/ Design Illustration

清雅含蓄、幽深清远、淡泊怡然的东方式神韵是设计师陈贻和张睦晨此次会所设计所要表达的和传统意义不同的东方风格，设计师将东方风格的自身形态、形式美感与精神意境通过自己的解读深入到创造的境界中去，为我们构筑出一个充满艺术灵性及自我感悟的空间设计作品，空间中自始至终渗透着"淡泊明志，宁静致远"的东方式精神内涵，同时又与中国传统"天人合一"的美学思想暗相契合。

通过使用当代文化的造型语言方式去寻求中国传统文化脉络，延续设计中的以意取象，诠释着设计师独特而又不可取代的"禅意"情绪。每一种视觉元素都在设计师精心安排下释放出特有的灵性与诗情，满足我们内心存在已久的审美需求。此套设计象征性地以符号化的探求向我们展示了更新的艺术格局，解构并重组了中式传统建筑中的"斗拱"结构，木制的桁架结构形成了空间中统一的造型元素而被巧妙地铺成开来，特殊的结构形式构成了独特的空间格局；或恢弘、或平和、或高耸、或延展，空间节奏处理得跌宕起伏、游刃有余。具体地说，在此套设计中将那种散点式的、全景性的、可居可游化的传统山水画艺术格局完全地融入整体空间构成中去了。设计中大胆地运用了白色抽象化的丁香花的视觉元素，让丁香花盛开在中央高耸挺拔的挑空空间，给人以一种新的审美意义的视觉形象，视觉上的冲击力加上东方式特有的艺术氛围让空间中的自然元素更加柔和地融合在了一起。

"心静、思远，志在千里"。无论从立意上，还是表现手法上，我们都能很好地感受到此空间巧妙地融合传统东方文化和时代精神。它既是一种传承，亦是一种升华；它既汲取了东方文化的传统意境精髓，又致力于发展和倡导优质的东方式生活理念及人居合一式的现代设计精神。

Elegant and implicit eastern style is all the designers pursue in this project. To achieve beauty that presented in external forms and internal artistic conception, they have put their interpretation towards her and built this artwork filling with the eastern aesthetics "live simple life, set your mind free".

Although she speaks a modern language, deep down her heart is traditional and poetic. To meet our requirements of beauty, every visual element that they meticulously put gives a new look to the space. For example, the conventional arch and wooden truss framework are transformed ingeniously into dramatic touch. The whole space develops scattered and systemic planning thoroughly. A white bold abstract lilac blooming high up the top of the ceiling impresses us with her eastern ambience and natural essence.

No matter from the conception or design approach, this space encompasses eastern culture and sense of contemporary times. More importantly, it even enhances an eastern living style.

323
IDEA-TOPS
艾特奖

NOMINEE FOR BEST DESIGN AWARD OF COMMERCIAL SPACE
最佳商业空间设计提名奖

深圳市中建南方建筑设计有限公司

获奖项目/Winning Project

博林天瑞售楼处
Bolin Tinery Sales Center

设计说明/ Design Illustration

博林天瑞毗邻深圳大学城，共享两大中心公园，选址于南临塘朗山原生山脉，北望西丽湖、高尔夫绿地、羊台山之地，具有独特的360度山湖林海四维一体生态景观视野，林荫大道丽水路都将成为其私家后花园。
本设计基于项目所在地生态景观的独特性和优越性为考量，以室内景观化为设计立足点，尝试营造外部景观资源在室内的延伸，以实现人文景观与自然景观在室内环境中的融合。
拟人的自然与自然的人境，是本项目要探寻的设计意义。这一核心设计观念贯穿在本项目的多个设计环节。如在功能分区和动线设计上，自然景观的自由和开放、简单和清晰，被转换成单一的曲线、曲面材质在场内的非常规的流动以及裂变，大坡道、平台、艺术长廊、沙盘展示、VIP室等功能空间化身为各个景观节点，由空间大曲线所统领，宛如自然生长，不突兀，不做作。而由高度立体化曲面所形成的多层功能空间布局，亦创建了欣赏场地全景的绝佳场所。
该项目设计灵感虽取自天然，但并不仅仅朴素，华丽而典雅的材质富于细节，在巨型水晶吊灯的映照下熠熠生辉，整个营销中心亦具有如豪华游轮般的奢华气质。
3000平方米，10米挑高，人与自然融合的语义在此有了充分且大气的表达空间，博林天瑞营销中心的震撼气势为客户提供了全新的视角和体验，项目品质感亦由设计细节所彰显。

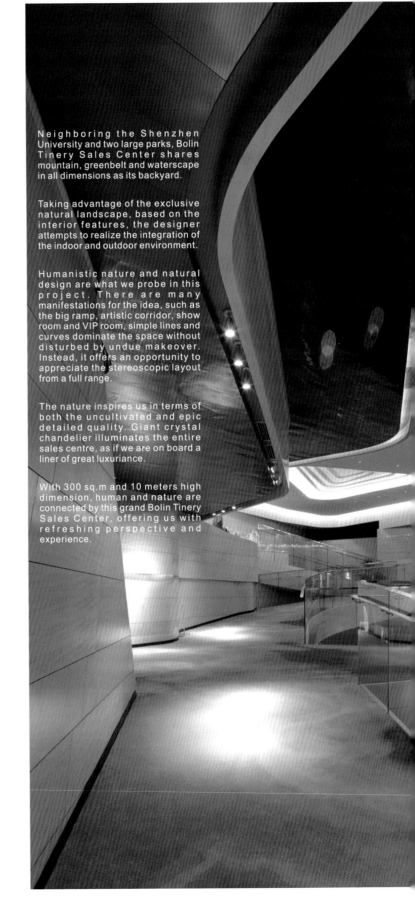

Neighboring the Shenzhen University and two large parks, Bolin Tinery Sales Center shares mountain, greenbelt and waterscape in all dimensions as its backyard.

Taking advantage of the exclusive natural landscape, based on the interior features, the designer attempts to realize the integration of the indoor and outdoor environment.

Humanistic nature and natural design are what we probe in this project. There are many manifestations for the idea, such as the big ramp, artistic corridor, show room and VIP room, simple lines and curves dominate the space without disturbed by undue makeover. Instead, it offers an opportunity to appreciate the stereoscopic layout from a full range.

The nature inspires us in terms of both the uncultivated and epic detailed quality. Giant crystal chandelier illuminates the entire sales centre, as if we are on board a liner of great luxuriance.

With 300 sq.m and 10 meters high dimension, human and nature are connected by this grand Bolin Tinery Sales Center, offering us with refreshing perspective and experience.

艾特奖
最佳光环境艺术设计大奖
BEST DESIGN AWARD
OF LIGHTING DESIGN

IDEA-TOPS
INTERNATIONAL
SPACE DESIGN
AWARD

329
IDEA-TOPS
艾特奖

获奖者
邵唯晏（中国·台湾）

获奖项目/Winning Project
YouBike-台东都历游客中心 / YouBike Tourist Center

获奖项目/Winning Project

YouBike-台东都历游客中心
YouBike Tourist Center

设计说明/ Design Illustration

台湾最后一块净土——美丽的台东
台东不论是感性的天然美景或是知性的人文生态，在地形上、生态上亦有其得天独厚之处，加上丰富的原住民、史前文化，共同织出东部海岸的迷人风华。而台东这所有的美好都能回归到最自然的三个元素——天、地、海，因而整个展览馆分为大地区、海洋区及天空区三大区，另外还有一个数位星空电影院。

东海岸风景区管理处的游客中心，定位上是给游客一个全新的角度来了解东海岸，带点教育但不冗长乏味，游戏化取代传统的展览陈设，互动的多媒体科技增添趣味性与国际性视野，动态与静态的展示结合就像这片东海岸，看似平静却处处富有生机，同时也是台湾首次大量将电脑参数式设计方法导入室内设计，利用大地、海洋及天空以大尺度折板系统来赋予空间戏剧性的张力。

全新的参观体验
脚踏车已经是近来节能环保的热门运动，我们将花东海岸自行车的动线导入，将游客中心转换为一座BICYCLE-FRIENDLY的展示馆，也是全台湾第一座可以骑脚踏进入室内参观的展览馆，不用担心车离身的烦恼，在展示馆中忽高忽低的遨游参观，是一种全新的参观体验。

电脑辅助设计
本案因为政府标案，在最低标的既有政策下，在有限的预算及工期下，要如期完成形体复杂的展示厅，相当困难。为克服以上的先天劣势，同时顾及工程品质，设计团队在设计前端采取大量的电脑辅助设计来介入整个流程。从设计的发想到施工图面的绘制，有效透过参数化的设计流程，以不同于以往的施工图呈现方式，终将本案执行完成。

The last pure land in Taiwan-Tai-Tung
No matter the emotional or intellectual human ecology, Tai-Tung has its unique view. She is rich in indigenous, prehistoric culture, weaving together a charming elegance east coast. According to these factors, we divided this exhibition into three parts: The Earth areas, Marine areas and the Sky area, as well as a digital stars cinema.

Visitors Center of East Coast National Scenic Area Administration is to give the visitors a new perspective to understand East Coast. Educational but not tedious, replacing the traditional exhibition with the game of interactive multimedia technology. Not only provides fun and joy but also enhance the vision of international. It is the first time to import a large number of computer parametric design method to design an inner space as well. The use of the flap system (folded plate system) gives the earth, the sea and the sky a dramatic tension.

The new tour experience
Bike has been a popular energy saving campaign recently, we import a bike movement inside the interior space, converts this visitor center to a BICYCLE-FRIENDLY showroom. It is also the first exhibition that allows the bike to ride around.

CAD/CAM system
The case is a government's bid. In accordance with the government "minimum price" principle, it was very difficult to finish the engineering work of the exhibition hall in time. Under limited budget and time, we applied CAD/CAM system in the process of design and construction. Differed from the past, parametric design process was fully utilized from scratch to construction drawing in this project and successfully completed it within the time frame.

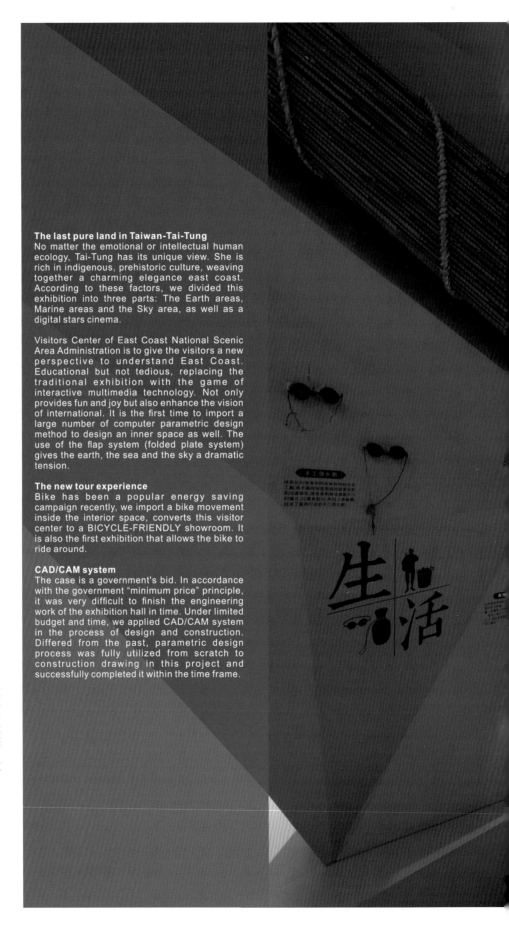

IDEA-TOPS
艾特奖

IDEA-TOPS
艾特奖

335
IDEA-TOPS
艾特奖

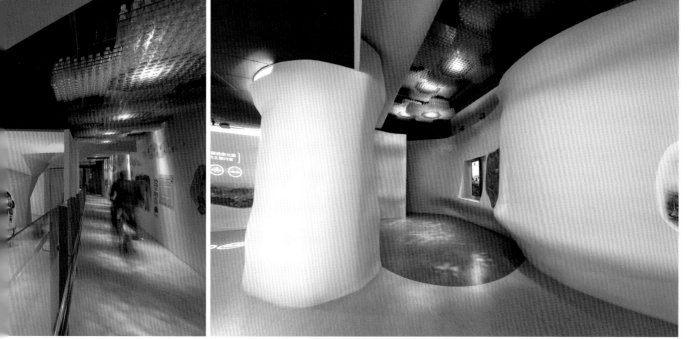

NOMINEE FOR BEST DESIGN AWARD OF LIGHTING DESIGN
最佳光环境艺术设计提名奖

浙江城建园林设计院有限公司光环境所

获奖项目/Winning Project

银川回乡文化园照明设计
Yinchuan Lighting Design

设计说明/ Design Illustration

银川回乡文化园项目总规划面积为19.5公顷，位于宁夏银川，是中国回族最大聚集地，一个有着浓郁回族风情的地方。银川回乡文化园是宁夏四大特色旅游之一，是宁夏唯一集中展示和弘扬回族民族文化的旅游景区。园中全国唯一的"回族博物馆"展示出回族的形成和发展历史。

全面采用黑钻系列LED泛光灯，类别为27W、60W、135W RGB。第一：原装进口CRE-E LED光源，提高光效，可达120lm/W；第二，鳍片式散热器结构，完美解决LED灯具散热问题；第三：与传统泛光灯相比，做R-GB/RGBW变色，不仅节能效果显著，而且使用寿命长，可瞬间开启。以"绚丽、夺目并表达文化"为出发点，在3D灯光秀的带领下整体建筑色彩变幻多姿，成了当地第一个夜间灯光"SHOW"场。采用LED幻彩变化、裸眼3D投影技术，为银川文博会带来一场光影变幻的视听盛宴。

The City of Hui Cultural Garden with 19.5 hectares in Yinchuan Ningxia Province, which is the biggest city for Hui religion, is one of the four feature gardens in Ningxia and is the only one tourist attraction for demonstrating the culture of Hui.

Introducing LED floodlights of dart diamond series at full scale, the imported CREE LED boost the luminous efficiency, the fin type radiator solves the cooling system perfectly and other merits includes energy saving, long service life and responsive to reaction.

Based on the concept of splendid, dazzling and cultural stance, it turns into a symphony of evening lighting Show place at the local. The LED Fantasia and EVO 3D finish enables Yinchuan a feast of audio visual.

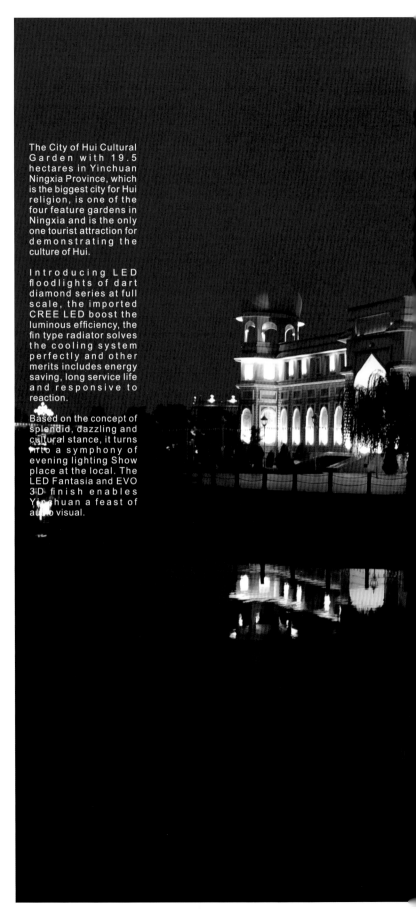

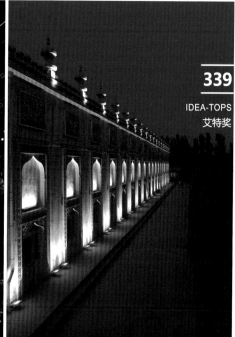

NOMINEE FOR BEST DESIGN AWARD OF LIGHTING DESIGN
最佳光环境艺术设计提名奖

罗伊真（中国·台湾）

获奖项目/Winning Project

台湾博物馆照明设计
Lighting Design of Taiwan Museum

设计说明/ Design Illustration

抓住绿色的脉搏，生命力的跳动从眼底逃脱，歌咏这城市美丽的诗颂！每一个城市都有属于自己的公园绿地，如同纽约的中央公园，已成立于1908年的台湾博物馆即位于台北市的二二八纪念公园，绿意林映，是台湾历史最悠久的博物馆，主要馆藏为台湾人文史料与自然产物。

每一建筑有其故事，漫步建筑，如同阅读一本好书般，细细品味，隐藏于建筑身后的历史文化、美学价值等多种特性融和之美。

台湾博物馆外观上糅合了多种古典西洋建筑元素，逐级而上的阶梯、巨大的希腊多立克柱式、饰有华丽花叶纹饰的山墙，搭配顶端的罗马穹窿圆顶，建筑风格具有文艺复兴时期希腊式神庙特色，望之庄严而神圣。不仅是日间，时至今日，台湾博物馆夜间风貌，因着照明设计上的层次对比、色彩变换，远比日间自然天光辉映下，更为雄伟、灿烂。

台湾博物馆具百年历史，基于对古迹维护的尊重，对老建筑一砖一瓦的保护，是面临整个照明设计案的艰难考验。

对建筑前方的柱列、壁柱、山墙，设计上采用以离地5.5公尺高度的远距立杆，架设全彩色系的LED投光灯，取代埋入式投射灯的方式，灯具立杆巧妙地隐藏于前方广场上进行投射，使往来行人不易察觉，再辅以前端挡板控制光源出射角度，以集中强化投射光束于柱面，并大大地降低对人与环境的干扰，巡礼在掩映着历史光影的环境中，感觉时间如长河般延伸。

当夜幕低垂时，华灯初上，透过光的语言及色温的层次，诉说老建筑的故事，藉暖色系2100K~3000K的色温呈现古意盎然，记住过去的时光，却不停留在过去。

适逢节庆等特定日时，将光艺术及色彩意象结合个别主题，透过DMX全彩色系的变化，注入艺术活力，充满魅力的光芒，独特韵致。例如：农历年节、馆庆(世界地球日)、情人节、中秋节及圣诞节，这仿佛将百年台湾博物馆换上了新娘的嫁衣，浪漫的粉紫色、纯净的靛蓝色、神往的琥珀色、晕眩的洋红色，风情万种，在夜空中熠熠生辉。

台湾博物馆的罗马穹窿圆顶，气势磅礴，宛如欧洲教堂的苍穹圆顶，分别以16盏不同角度的LED灯投光灯，陈列安装于圆顶及两侧区域，投射圆顶四周，进行同步的色温及彩度变化，方圆互蕴光影相映，与周遭光环境协调共存。月上柳梢头，光泽匀称的罗马圆顶，使来到台北车站的旅客抬头远望，清晰可见而深刻地触及人心，耸立参天，幽情缤绚。

在前方广场的立杆上，安装透过特殊滤片的复金属灯，使光源精准投射山墙上的雕刻纹路，在LED灯的光影变换间，呈现渐层、柔和的气氛，勾勒出墙上丰富华美的山花叶纹饰雕刻，建筑的古意与光影的精致相互交融，使伫立的行人不时发出视觉上的惊叹。

在主建筑两侧廊道壁柱，以琥珀色系投光灯上照洗练，将漫射的光刻画出一丝一缕的历史痕迹，在夜间展现悠然沉静的氛围。

观察台北城市核心的发展，从早期的艋舺、大稻埕，位于台北市西区，迄今转移至位于台北市东区的信义计画区。随着邻近台北车站双子星大楼的规划、兴建，可以预见台北都市的核心即将由东区转移至西区，换上嫁衣的台湾博物馆，与众不同的夜间风貌，百年荣耀，风华再现，继续述说台湾之美。

Green is her life and pulse in praise of our beloved city. Every city has her garden and museum. Taiwan Museum was found in 1908 in Taipei, cultural chronicles and products of the nature take an important role in the collection.

Just as every building has her story, there are many history, culture and art buried in it as well.

The exterior of Taiwan Museum is a complex of western architecture, for instance, the ascending stairs, Doric columns of Greek, gorgeous embossed gables, round arches of Roman style and Greek temple during the Renaissance, so solemn and sacred. At night, it appears more magnificent and splendid than the broad daytime.

Due to the architectural heritage conservation matters, it is a great challenge to develop lighting design for the hundred-year old Taiwan Museum.

Considering the pillars, columns and gables on the facade, it applies full color standing floodlights instead of stowed spotlights. Besides, the standpoint is controllable focusing on the façade so as to minimize interference to people and environment.

When evening falls and the lights on, this historical building tells stories through its changing of color tones, memorizing the old days while not be stuck in the past.

When festivals come, lighting artistry will be rendered with symbolic charm as if it changes clothes of violet, indigo, amber and magenta, glittering in the dark backdrop.

The Roman style dome of the museum shone by 16 pieces of LED floodlights that displayed at the top and two sides fits in the surroundings. Usually, this spectacle impresses the travelers who arrived at Taipei Station in no time. Square space at the front where pole lamps are set, special filters are installed on metal halide lamps producing ambience of multiple light and shade. At the sight of old building in renewed looks, passers-by will stop and.

Pilasters in corridors by both sides are projected by amber colored lights, rendering historical ambience and silent night.

Monga and Twantutia are in the west of Taipei which were the city centers back to early stages, while at today it shifts the core of development to the east. However, as more and more buildings rising in the west along with the renewed Taiwan Museum, a more prosperous era is around the corner.

IDEA-TOPS
艾特奖

NOMINEE FOR BEST DESIGN AWARD OF LIGHTING DESIGN
最佳光环境艺术设计提名奖

北京八番竹照明设计有限公司

获奖项目/Winning Project

重庆云会会所
Chongqing Cloud Club

设计说明/Design Illustration

灯光设计构思：云会会所位于重庆的龙兴古镇。古镇有着数百年的文化传承，浓郁的川蜀文化根深蒂固。成功的照明光环境设计不是越张扬越好，恰恰相反，柔和、隐蔽、色彩统一的光更能体现和表现设计。如何与环境相结合，避免眩光，巧用设计手法，而又不被环境所完全淹没。会馆餐厅的整体风格及氛围非常和谐统一，经过精心布光设计后，会所内外，从景观—建筑—室内，灯光表现得都依从于介质本身的需要进行秩序化梳理，从而达到高雅、精致、从容的光感受。

Cloud Club locates at the centuries old Longxing town of Chongqing city, for which town is rich in Sichuan culture. The key to a successful lighting design for us is nothing flashy but soft, concealed and well-conceived one with integrated color and finish scheme. Therefore, how to cohere with the environment and avoid discomfort glare have been our primary concerns. The style and ambience of the Club is harmonious and unitive from landscape, exterior to its interior, the lighting is sculpted in compliance with each feature by good order to attain a refined, exquisite and calm sense.

NOMINEE FOR BEST DESIGN AWARD OF LIGHTING DESIGN
最佳光环境艺术设计提名奖

关永权照明设计（北京）有限公司

获奖项目/Winning Project

小南国花园酒店
Shanghai Min Gardern Hotel

设计说明/ Design Illustration

小南国花园酒店是由小南国集团酒店管理公司经营管理的第一家都市度假酒店，隶属于小南国集团。室内设计由香港BLD设计公司设计，灯光由关永权照明设计（北京）有限公司承接。

酒店坐拥上海杨浦区的核心地带，俯瞰黄兴公园的绿意盎然，与碧波荡漾的湖景交相辉映的小南国花园酒店楼高26层，166间客房均面湖而设，宽敞的落地景观窗将黄兴公园的绿意盎然的景致尽收眼底。酒店邻近多家知名高等学府，步行5分钟即可到达黄兴公园地铁站，无论是来参加婚礼的宾客们还是前来会晤的商务人士，不用"踏破铁鞋"便能寻见。

由于小南国酒店在设计期间正是LED开始大量用于室内设计阶段。在该项目中，全部暗藏灯槽都使用暖光2400K LED灯带，为酒店间接光提供柔和光效，同时为酒店后期运营节约能源。灯光设计还是关永权照明设计（北京）有限公司一贯秉承的"运用最少量的灯具营造最好的灯光环境"。在需要表达室内设计精彩的地方，设计师不遗余力地用了足够的照明灯具来表现。在不需要的地方，设计师很吝啬地布置每一个灯位。要求每一个灯位都有它实际存在的意义。

酒店拥有可容纳90位宾客的礼堂，许许多多新人们留下了他们约定终生的誓言。落日西沉，结束了梦幻的婚礼仪式，柔美的灯光从教堂入口两边水池池底投射至教堂外墙，亲朋好友们移步至与婚礼教堂相连的600平方米的露天平台，尽情融入西式婚礼的时尚派对，为新人们送上最真挚的祝福。教堂内部，灯光赋予了建筑本身的特色，暗藏的暖光灯管，勾画出建筑轮廓，LED埋地灯从建筑底部安装，更好地表达了建筑向上凝聚的弧形。从入口处开始，沿步道位置布置埋地灯，点点灯光引导新人走上神圣殿堂。

位于酒店3层的AdD是一家环境舒适悠闲的全日制餐厅，暗藏灯带提供了间接照明，布菲台和桌台用高显色性光源，提供客人最佳就餐环境。

酒店的钻石宴会厅是浦西市区最大宴会厅之一，可同时容纳90桌宴席，面积达1200平方米；宴会厅铁艺镂空雕花背景墙，采用洗墙灯的照明方式，把宴会背景照亮，入口处竖向艺术玻璃，采用背光方式，提供宴会厅垂直面的照明。

酒店飞SPA中心面积达5000平方米，覆盖酒店副楼4个楼层，拥有热疗区，还设有9间位于主楼的水疗理疗包房。飞SPA拥有22间豪华水疗套房，客人可以直接入住，并在私密的个人空间里享受各式水疗和按摩服务。从客人步入客房走道的瞬间就会感受到安静的氛围，暗藏灯槽提供了舒缓的过度空间的光氛围，房间入口处用侧壁安装的嵌入式灯具提供了安全指示功能。让整个SPA空间温情舒缓。

酒店用独特的声光影像科技搭建了一条"食"光隧道，用声画重现了清朝贵族食坊、民国百乐门舞厅、弄堂石库门、"文革"时期的工厂食堂以及今天的浦江风情到未来"绿色"空间等场景，以"饮食"为主题演绎出一场年代更迭的上海往事。灯光也赋予这几个不同的餐饮空间不同的灯光氛围。从灯光昏暗柔和的百乐门到高色温光环境的工厂食堂，再到梦幻的灯光营造的飞SPA中心。飞SPA食光隧道与飞SPA中心组成了上海小南国花园酒店内的一个休闲养生的综合区域，是一个动静相宜、灵活舒适的体验馆。

Shanghai Min Garden Hotel is the first urban resort hotel of Xiao Nan Guo Group. The interior is designed by BLD of Hong Kong whilst lighting is designed by Tino Kwan Lighting Consultant (Beijing).

Situated at the core of Yangpu district of Shanghai, it is a 26-storey building that offers 166 hotel rooms with spectacular lake view overlooking the lush greenery in Huangxing Garden, adjoining several preeminent universities and Huangxing Garden metro station. Therefore, no matter you are a wedding guest or businessman, it is bound to be an appropriate alternative for you.

As LED is widely applied and effective in interior designing, all the hidden lamp slots in this project utilize 2400K LED as indirect illumination plus energy saving all at once. It is also the philosophy which Tino Kwan Lighting always conceives that "Minimal lighting equipment to achieve maximum lighting effect." The designer spares no efforts to dispose every light to its place, and each light counts.

Ballroom accommodates up to 90 guests imprinting many newly-weds with precious memory. In the evening, graceful lights from the bottom of the pond on both sides by church entrance are projected to the exterior wall, connecting to an open terrace of 600 sqm where western wedding party puts on. Hidden warm lights fixed indoors draw the arcs and lines of the church. Starting off the entrance, sparkling lights along the path lead newly-weds to the hallowed auditorium.

AdD, a cozy restaurant on third floor, offers full-day service to the customers. Buffet and bar tables provide guests an optimal dinning space in high color rendering scheme.

With more than 1,200 sqm of meeting space, it is one of the biggest grand ballrooms in Puxi district that can hold up to 90 banquets. The backdrop wall with metal cut-out design is illuminated by wall wash while the entrance applies back lighting system.

The Fei SPA features a space of 5,000 sqm and covers 4 storey including thermotherapy area and 9 private hydrotherapy rooms. There are 22 luxury suites with hydrotherapy for customer to enjoy a private service. The hallway to the room features a peaceful ambience by hidden lights.

Sound, light and digital image in the hotel build a time machine for themed restaurants of Shanghai, such as Qing Dynasty, Republican time, Shikumen style, Cultural Revolution time and futuristic green space. They are either bathed in dim and soft or high color temperature luminaire. The magical Fei SPA and Times Restaurant, the health and recreation section are an integrated part in of Shanghai Min Garden Hotel.

简一 大理石瓷砖
因为专注　所以更逼真

大理石的逼真效果
瓷砖的优越性能

| 大理石瓷砖　🔍 搜索 Search |

4006-907-908

Stonelution
施朗格·石砖

施朗格
科技让大理石更完美

TECHNOLOGY ALLOWS
MORE PERFECT MARBLE

佛山市施朗格石材有限公司　　佛山市季华西路陶瓷总部基地中区C06　　TEL:400 622 1699

深圳国际家居软装博览会

3.7-10 8.7-9

同期举办：家居中国（深圳）创意设计周

展览地点：深圳会展中心 全馆盛大开展

扫一扫，快速预登记

主办单位：
全国工商联纺织商会.
广东省家用纺织品行业协会
Andrew Martin 奖中国区总代理.
深圳室内设计师协会

I WILL BE ALWITH WITH THE MASTER
我与设计大师同在

世界著名建筑大师扎哈·哈迪德在上海的旷世杰作，每束灯光都由设计师精心调配设计，**凌空**
西顿照明，用灯光赋予建筑最美灵魂　SOHO

广东省惠州市惠城区水口东江工业区祥和西路A17号　电话：0752-5333999　传真：0752-53339999　西顿官网：WWW.CDN.CC

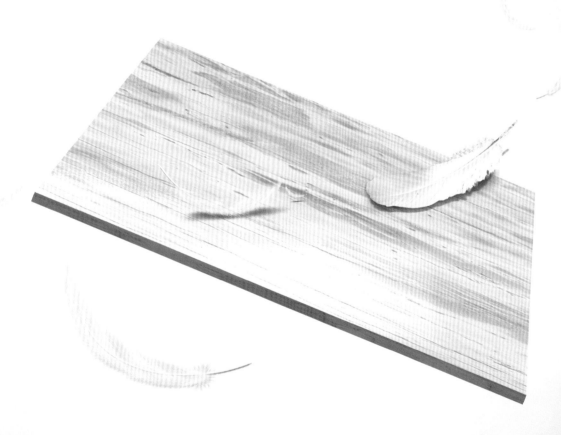

上国石材
TOPPING STONE

企业介绍
Company Introduction

北京上国石材有限公司成立于2008年，是土耳其CTM中土矿业的中国管理总部，拥有雄厚资金链以及独立运营资本。CTM中土矿业在土耳其拥有十多座矿山资源，目前已顺利开采2个品种，矿石资源已经运抵国内进行销售。开采出来的大理石产品，如摩根世家、苏格兰灰等，自入市以来即得到业界广泛好评，先后用于天津津湾广场、乌海市博物馆、成都温江天莱国际大酒店等著名工程。

公司致力于打造海内外石材矿业与中国设计互动的生态平台。以北京为总部，以打造精品工程为目标，辐射至香港、广东、福建、上海、沈阳，与多家大型国内外房地产企业形成战略合作伙伴关系，其中最具代表性的大型地标性项目如：上海汤臣集团合作的上海汤臣一品、天津的汤臣津湾一品等，与香港太古集团珠连璧合打造了广州太古汇广场和上海大中里太古汇；与保利地产共创了保利中央公园、沈阳保利康桥、海上五月花等。

公司长期专注于耶路撒冷金、贝砂金石材品种的开发和经营，该品种占中国销售份额的70%以上。2010年扩展至土耳其大理石矿石开采，拥有多座自有矿山，年产大理石方料超过20万吨，产品销往中国、欧美、中东等国家。公司拥有约30000平方米加工厂，从矿山源头选料到加工厂，可为大型石材装饰项目提供强有力的一条龙的保障服务。

上国石材自成立之初，便开创石材行业新篇章。开采经典石材产品是上国人一直坚行的路。铸就国际化市场、国际化人才、国际化管理，继往开来，上国石材期待与优质客户合作共赢，实现石材产业链的生态之梦。

咨询热线：400-819-1595
公司地址：北京朝阳区豆各庄乡马家湾21号 | 网址：www.toppingstone.com

上国官方网站

上国官方微信

10
在土耳其
拥有10座矿山

30000
在北京拥有30000
平方米加工厂

上国石材
我们拥有自己的矿山资源

尚品美居中国控股有限公司
SHANGPINMEIJU CHINA HODING CO.,LTD

赢在软装
WIN IN DECOR DESIGN

打造中国顶级营销型软装设计师，解密您心中的客户和客户心中的您！
引爆软装行业培训巅峰,高规格,高标准,高素质,国际导师亲临传授。

色彩运用、风格实践
家具氛围、嗅觉设计
风水讲解、花艺设计
消费心理、灯光布控
插花设计、布艺设计

如果三天三夜可以快速提升创作格局
如果三天三夜可以从设计到懂创作到形成作品
如果三天三夜可以从空间概念细化到花色风水
如果三天三夜可以和国际顶级大师碰撞智慧
如果三天三夜可以懂营销懂推广更懂软装设计
一定是在《赢在软装》

摆场经典 5

设计品味 4

绝对成交 1

完美陈设 2

关注微信　分享更多

尚品美居旗下精品《赢在软装》课程隆重开讲，这是刷新软装培训行业新标杆的新潮课程，是专为中国空间设计师量身定制的精品课程，是立志打造中国顶级营销型软装设计师的实战课程。《赢在软装》不仅仅是软装培训课程，也是学习如何推广自己，如何快速成交，如何塑造价值，如何掌握消费者心理的课程，在这里只要你带着发现的眼睛，相信一定会让你大有收获。《赢在软装》课堂不仅仅是学习的地方，也是资源分享，信息并存的场所，更是会给你一把打开财富大门的金钥匙，只要你来，只要你够诚恳，一定不会让你失望！

顶级师资 3

全国统一服务热线：**400-808-3592**

地址：广东省东莞市厚街镇家具大道29号柒号馆
Add: Furniture Avenue 29, Houjie Town, Dongguan, Guangdong
电话Tel/传真Fax：0769-85829558　QQ:327330376
邮箱Email：shangpinmeiju@163.com

奥莱·尼思 OLANIES

领袖级水晶灯

为 荣 耀 加 冕

OLANIES（奥莱·尼思）的诞生正是水晶灯历史上一个不朽的传奇。沿袭着纯正而悠远的欧洲王室血统，奥莱·尼思水晶灯在历史的长河中，一路闪耀着无上的君王之光。从摇曳多姿的塞纳河到大气沉雄的白金汉宫，奥莱·尼思浸染了太多中世纪欧洲皇族文化的精髓，从来都只为那些最炫目的荣耀而绽放光华！

公元2005年，宝辉国际灯饰集团正式将这一光耀全球的水晶灯品牌—OLANIES（奥莱·尼思）推向中国大陆市场，同年亦成为施华洛世奇®元素特约授权品牌。短短几年内，奥莱·尼思迅速以其高贵典雅的设计、精湛绝伦的工艺、顶级奢华的材质获得高端人群的认同和礼赞。奥莱·尼思卓尔不群、耀而不奢的领袖气质，为那些在事业和生活各方面已赢得至高荣耀的人们，辉映出最闪亮的光芒！

www.olanies.net.cn

奥莱·尼思全国经销商

·北京 ·上海 ·广州 ·深圳 ·重庆 ·东莞 ·长沙 ·武汉 ·南京 ·昆明 ·太原 ·天津 ·沈阳 ·厦门 ·西安 ·宁波 ·杭州 ·郑州 ·义乌 ·绍兴 ·无锡 ·温州 ·东阳 ·常州 ·昆山 ·余姚 ·宜昌 ·福州 ·福清 ·泉州 ·普宁 ·兰州 ·大同 ·孝义 ·台州 ·慈溪 ·盐城

集团总部： 香港九龙旺角亚皆老街8号朗豪坊办公大楼3715室
中国品牌营销中心： 广东省中山市古镇镇中兴大道星光联盟五楼C区01-12

香港寶輝國際燈飾集團·榮譽出品
THE WORLD'S LEADING LIGHTING SPECIALISTS

威伦斯
RENAISSANCE

人文艺术水晶灯
人文之光　传世百年

威伦斯（Renaissance），宝辉国际灯饰集团有限公司旗下品牌，
一个设计灵感源于欧洲文艺复兴艺术天堂佛罗伦萨的灯饰品牌。
威伦斯（Renaissance）背后，有着深厚的欧洲文化底蕴，
尤其是文艺复兴时期"人文主义"的思想浸润。
意大利的浪漫、文艺复兴时的文化艺术精华共同构筑了威伦斯（Renaissance）的文化基因，
使其成为人文艺术水晶灯流派的代表之作。

威伦斯全国经销商

·北京 ·上海 ·广州 ·重庆 ·天津 ·沈阳 ·杭州 ·太原 ·厦门 ·长沙 ·惠州 ·佛山 ·成都 ·韶关 ·长春 ·郑州 ·武汉 ·青岛 ·齐齐哈尔 ·西安 ·吉林 ·宁波 ·台州 ·无锡 ·盐城 ·余姚 ·常州 ·南通 ·温州 ·富阳 ·慈溪 ·昆山 ·义乌 ·昆明 ·石家庄 ·荆门 ·衡阳 ·福州 ·丹东 ·大同

集团总部： 香港九龙旺角亚皆老街8号朗豪坊办公大楼3715室
中国品牌营销中心： 广东省中山市古镇镇中兴大道星光联盟五楼C区01-12

香港寶輝國際燈飾集團·榮譽出品
THE WORLD'S LEADING LIGHTING SPECIALISTS

PERFECT STONE MODEL
FIVE STAR STONE KINGDOM

完美石材典范
打造五星级石材王国

HC.STONE

皇朝石材集团是最早实现石材集成、一站式采购的终端服务商，历经十余年的快速稳步发展，已成为集矿山开采、销售设计、大型工程承接、多品类石材生产加工、安装维护为一体的多元化、跨地域的企业。

皇朝石材集团以先进的加工设备、精湛的技术工艺、高效的服务团队为我们的高品质客户提供最优质的产品以及最满意的服务，完美展现石材独有的原创和独特的装饰效果，缔造奢华夺目的艺术空间。

皇朝石材集团标志工程：深圳T3航站楼四层安检大厅，深圳T3航站楼平安银行VIP接待厅，卓越时代广场，深圳罗湖边检出入境大厦，中国银行蛇口鹏龙支行，工商银行宝安丽景支行，平安银行景田支行，万科武汉君澜酒店，万科双月湾，万达西双版纳文华酒店，万达成都瑞华酒店，万达苏州嘉华酒店，武汉白玫瑰酒店，珠海棕泉酒店，佳兆业大都会可域酒店，深圳回品酒店，广西北海富丽华大酒店，金积嘉集团万国食品城，华南城总部大楼，华南城发展中心，华南城东北城总部，香港落禾沙项目，保利地产广州琶洲项目，深圳远鹏装饰大楼，高邮陆宇中央郡凤凰城别墅，招商地产伍兹公寓，深圳南岭一半山，新世界地产世纪御园，湖北宜昌恒信中央公园，沈阳紫薇仙庄，安徽徽商集团酒楼项目，海底捞广州店，海底捞深圳店及深圳高端私人别墅等。

务实创新是皇朝石材集团坚定的发展方向。坚持以客户满意为立足点和出发点，以高效服务、品质至上为工作重心，坚定不移的走品牌化发展路线。创新是企业发展的不竭动力。在务实发展的同时，与时俱进，积极研发独特的技术工艺与营销模式，深度理解和发掘"石文化"内涵，积极探索石材装饰行业发展新趋势，独创石材商业运营新模式。

以卓越的供应与服务，成为全球石材行业的先导。皇朝集团将以石尚艺术为帆，以与时俱进、务实创新为力，引领石材行业发展新风尚。

香港皇朝石材集团控股有限公司
广东皇朝石材工艺有限公司
深圳市皇朝石材有限公司
深圳市龙岗区南湾布澜大道盛宝路皇朝产业园
TEL:86-755-8996-3888　FAX:86-755-8951-2555
WWW.huangchaostone.com

尊贵服务热线：
13823237666 / 13026626666